LECTURE NOTES ON COMPUTATIONAL MUTATION

LECTURE NOTES ON COMPUTATIONAL MUTATION

GUANG WU AND SHAOMIN YAN

Nova Science Publishers, Inc.

New York

For permission to use material from this book please contact us:
Telephone 631-231-7269; Fax 631-231-8175
Web Site: http://www.novapublishers.com

NOTICE TO THE READER

The Publisher has taken reasonable care in the preparation of this book, but makes no expressed or implied warranty of any kind and assumes no responsibility for any errors or omissions. No liability is assumed for incidental or consequential damages in connection with or arising out of information contained in this book. The Publisher shall not be liable for any special, consequential, or exemplary damages resulting, in whole or in part, from the readers' use of, or reliance upon, this material. Any parts of this book based on government reports are so indicated and copyright is claimed for those parts to the extent applicable to compilations of such works.

Independent verification should be sought for any data, advice or recommendations contained in this book. In addition, no responsibility is assumed by the publisher for any injury and/or damage to persons or property arising from any methods, products, instructions, ideas or otherwise contained in this publication.

This publication is designed to provide accurate and authoritative information with regard to the subject matter covered herein. It is sold with the clear understanding that the Publisher is not engaged in rendering legal or any other professional services. If legal or any other expert assistance is required, the services of a competent person should be sought. FROM A DECLARATION OF PARTICIPANTS JOINTLY ADOPTED BY A COMMITTEE OF THE AMERICAN BAR ASSOCIATION AND A COMMITTEE OF PUBLISHERS.

Library of Congress Cataloging-in-Publication Data

Wu, Guang, 1962-
 Lecture notes on computational mutation / Guang Wu and Shaomin Yan, authors.
   p. ; cm.
 Includes index.
 ISBN 978-1-60456-516-4 (hardcover)
 1. Mutation (Biology)--Mathematical models. 2. Nucleotide sequence. 3. Amino acid sequence.
I. Yan, Shaomin. II. Title.
 [DNLM: 1. Amino Acids--genetics. 2. Computational Biology--methods. 3. DNA Mutational
Analysis. 4. Mutation. 5. Sequence Analysis, Protein. QU 60 W965L 2008]
 QH461.W8 2008
 576.5'49--dc22
                              2008012727

Published by Nova Science Publishers, Inc. ≃ New York

Contents

Preface

Research economics. From 1994 to 1999 the authors were in depressed moodswhen they were working and studying in University of Udine Medical School, Udine, Italy. The situation was desperate because western researchersgenerally begin their scientific publications around 25-years old. With the speed of two publications per year, an ordinary western researcher would have 20 to 30 papers in international peer-reviewed journals at the age of 40. By clear contrast, the first author published his first full paper in 1995 at age of 39. The situation was not only miserable but also hopeless.

Thus, both authors concentrated all of their efforts on the issue of how to rapidly increase the number of their publications in order to reach the standard for western researchers.

At that time, two issues closely related to scientific publications drew the authors' special attention. The first issue was the impact factor. The impact factor in journals of pharmacokinetics, where the first author was working, generally ranged from less than unity to five [1], whereas the impact factor in journals of molecular biology was much higher, as high as 38 for the journal "Cell" in 1997 and 1998 [2].

This led the authors to have the following meditation. According to the impact factor, pharmacokinetics was an iron mine while molecular biology was a gold mine. People who own an iron mine can make more money than the people who own a gold mine, if the production of iron is incomparably larger than that of gold, for example, the production of millions of tons of iron versus the production of several grams of gold. However, in this sense, the researchers working on pharmacokinetics would absolutely be losers if they published their research papers at the same speed as the researchers in molecular biology, which is similar to the situation where the iron mine would produce several grams of iron per year.

The second issue was that the authors had nothing to write although they have had written several theoretical papers. In fact, the key point was that the authors did not have a lot of data to write papers. Another consideration conducted during those unforgettable days was as follows. For a researcher, his/her market is scientific journals, his/her factory is his/her educated skill, and his/her materials are the data. In contrast to the massive industrial production, where a factory can easily buy raw materials, the researchers must "dig" materials themselves such as collection of blood samples, biopsy specimens, etc. In fact, researchers spend far more time in collecting their materials than working in their factory, which is their educated skill. Sadly, an industrial factory can afford high prices for materials because the factory can raise the price of its product accordingly, whereas the researchers cannot afford the high price of materials because the products of researchers are not priced.

The only valid conclusion drawn from these two issues is that the authors must have unlimited, cheap (even free) resources in a highly impacted field if they want to dramatically increase the number of their publications and impact factor. The openly accessible data banks on DNA/RNA/protein are clearly the unlimited and free materials, with which the authors can produce as many papers as possible if they have a suitable factory to process these materials. Therefore, we invested our spare time and energy to build such a "factory" to process the protein sequences in 1999, which was the beginning of *Computational Mutation* although we did not realize it at that time with only a goal of increasing our number of publications.

Living and dynamic measures. Since then, we have significantly developed the computational mutation, and have figured out what are the key points in this new discipline. Now we would like to say that the key point in applying the quantitative methods to biomedical sciences may not be as simple as just taking the measures developed by physicists and chemists, who spent years to figure out the meanings of measures. We may need to define the real measures belonging to organisms and then apply them to quantitative methods.

Dimension. For the development of measures, an important issue ignored in biomedical sciences is the issue of dimension or unit. This is because the research we conduct is to establish some relationship between related events, with such a relationship we can predict what will happen in the future and explain what has occurred in the past. Naturally, this type of relationship is more likely to be the cause-consequence relationship, which however requires the dimensions to be eventually equal between the cause and consequence [3]. Otherwise, the relationship would be a phenomenological or empirical relationship, whose meaning would be somewhat limited. Our measures developed in this book are dimensionless, because we compute the probability of events.

Meaning of our study. The most important meaning for our work in past eight years is that we lay the foundation for the application of various mathematical models into protein, RNA and DNA studies because we defined three living quantifications to measure the dynamic state of proteins, RNA and DNA.

Readership. In general, we follow such a line to develop our approach: (i) we develop a new measure to quantify randomness within a protein, (ii) we observe the behavior of the new measure in spatial and time scale with real-life cases, and (iii) we apply the mathematical models to solving more complicated real-life problems.

Basically this book can be stratified into three parts: the first part includes Chapters 2 to 4, the second part includes Chapters 5 to 7, and the third part includes Chapters 8 and 9. Each part develops along the three-step frame defined above.

As the book is written from theory to practice, it generally can be used to teach the students without any experience in the laboratory. Although this book deals with statistics for comparison and model development, the students without training in statistics can also benefit from this book. The essence of this book is multi- and inter-disciplinary, therefore it can be used for the biomedical and non-biomedical students.

The aim of this book is to solve the very practical problem, that is, to predict mutation at the amino-acid level and at the RNA codon level. A vast majority of examples in the book are related to the surface proteins, hemagglutinin and neuraminidase, from influenza A viruses, so the researchers working on this topic can find their interests in this book. In addition, the methods in Chapters 10 to 12 can be applied to any proteins to conduct a formal research project.

Guang Wu, MD, PhD and Shaomin Yan, MD, PhD
December 25, 2007
Yangshuo
Guang Xi
China

Chapter 1

Measuring Living Protein

A highly abstract knowledge in humans is mathematics, which unfortunately has only limited applications in biomedical fields although it is being very successfully used in other fields, such as physics and engineering. This is by no means because the scientists are not interested in applying the mathematics to biomedical fields. By contrast, tens of thousands of scientists around world are making great efforts to apply mathematics to the biomedical field [4, 5].

One question raised here is why we, scientists in biomedical fields, need mathematics. In fact, we have used a lot of mathematics in biomedical fields, for example, we use statistics to analyze the experimental data [6, 7], we use pharmacokinetics to analyze the blood drug concentration [8, 9], etc.

However, the most important aspect of mathematics is yet widely to be used in biomedical fields, that is, the predictability of mathematics, with which we can send the humans to moon, for example. The mathematical predictability comes from mathematical modeling, a relationship between events of interests.

An interesting area that we may apply a lot of mathematics is DNA/RNA/proteins, of which the data are currently accumulating exponentially [10]. Actually, a lot of mathematical tools in conjunction with computer are involved in the analysis of DNA/RNA/protein data [11-13].

With respect to DNA/RNA/protein structure, an important issue is to predict the mutations, however, this has yet to be solved. We will address the prediction of mutations in the primary structure of a protein in this book.

The accurate, precise and reliable prediction comes from the suitably mathematical models. As we just mentioned, mathematical modeling is a relationship between events of interests. For example, we might be interested in when a mutation will occur, i.e. we are interested in the relationship between mutation and time. Now let us look at this relationship, where we only need two elements, mutation and time. Things are quite simple! However, until now we have no reliable predictions on when a mutation will occur, so the only logic and convincing explanation in such a relationship is too complicated to be solved even though there are only two elements.

Still we might be interested in which position a protein sequence mutation will occur, i.e. we are interested in the relationship between mutation position and X, Y, and Z. So what are

X, Y and Z in this relationship, honestly we do not know, but only one thing is sure, say, X or Y or Z must be some measurements in proteins. This is understandable, because we need to feed any mathematical model with numbers.

For the question of X, Y and Z, let us see what measurements we have with respect to a protein. We can measure the length of a protein, i.e. a protein contains how many amino acids, we can measure the composition of a protein, i.e. the contribution of 20 kinds of amino acids to a protein, and we can measure some parameters at high-level structures. However, these seem to have little help for building a relationship between mutation positions and X, Y and Z because a mutation can occur at any position whereas these measures are not related to the concrete position in a protein.

Of course, researchers have been trying to use various measures, which are related to each amino acid, in different analyses [14, 15], but not in predictions of mutation! Actually, it is very easy to think out that we can apply the physico-chemical property to each amino acid, which is indirectly related to the position in a protein. However, the biggest problem is that physico-chemical properties as well as other measures are "dead" measurements, i.e. they do not present the living state of a protein, for example, the solubility of alanine at 25 °C is 16.65 g/100g, its crystal density is 1.401 g/ml, its pI at 25 °C is 6.107 and so on. However, these properties do not change when the alanine locates at different positions of a protein.

This nevertheless is the biggest problem we meet in trying to apply the mathematical models to studying many biomedical phenomena. The things we study are alive, but the measures we use are dead. We frequently hear the complaints made by bio-physicists that what they measure is different at different situations!

All of these mean that the DNA/RNA/protein science strongly needs the new methods, with which we can measure the living DNA/RNA/protein. Historically, this marks the development of a scientific discipline reaching a new level, for example, the development of geometry originated from the measurement of land [16].

Developing the methods to measure a living protein to the scale of each amino acid is our main purpose for most parts of this book. With quantified living protein, we can build the relationship to couple the mutations with X, Y, and Z, for example.

This aim seems to have somewhat spiritual characteristic, as we cannot feel and touch it physically, but it must be alive and subject to any change in a protein. Actually, randomness does have such a "spiritual" characteristic, because pure chance is now considered to lie at the very heart of nature [17]. "Random" and "randomness" have been defined as follows in dictionaries, for example, (1) without pattern: done, chosen, or occurring without an identifiable pattern, plan, system, or connection; (2) lacking regularity: with a pattern or in sizes that are not uniform or regular; (3) statistics equally likely: relating or belonging to a set in which all the members have the same probability of occurrence; (4) statistics having definite probability: relating to or involving variables that have undetermined value but definite probability [18]. Another example of definition of random is as follows: random, adjective, date: 1632, 1 a: lacking a definite plan, purpose, or pattern, b: made, done, or chosen at random; 2 a: relating to, having, or being elements or events with definite probability of occurrence, b: being or relating to a set or to an element of a set each of whose elements has equal probability of occurrence [19].

These definitions do seem to have "spiritual" characteristics, however, we can simply understand that the meaning of randomness in our sense is that an amino acid with a bigger mutation probability would more easily mutate than an amino acid with a smaller mutation probability in a protein. Thus, the question here is how to find a measure to attach the mutation probability to each amino acid in a protein.

Besides, randomness also suggests the parsimony of nature, that is, nature will use the least time and energy to do the things such as construction of proteins.

This way, randomness may not be as empty as we initially thought. In the following Chapters we will define and quantify randomness in a protein, and observe the behavior of quantified randomness along spatial and time axes, and apply the quantified randomness to mathematical models.

Chapter 2

Amino-Acid Pair Predictability

It is well known that a protein sequence contains information, which can be measured by Shannon's entropy [20], but this only presents the view in terms of Shannon's entropy. Actually, we may have several ways to read the information contained in a protein. However, we generally do not know how to read this information correctly because we do not know (i) whether or not a single amino-acid "word/letter" is constructed by three amino acids as that a single DNA or RNA "word/letter" is constructed by three codes, (ii) how many amino acids construct a single amino-acid "word/letter", and (iii) whether or not there are "punctuation" and "space" in a protein sequence.

The simplest way to analyze the meaning of an unknown language is to count the frequency of each element, and then we compare the counted frequency with that in a known language. For example, the letter "e" appears most frequently in English [21], thus the most appeared element in an unknown language is highly likely to correspond to the letter "e", if we do not consider the modern encryption technology, which deliberately changes the frequency.

2.1. Counting Amino-Acid Sequences

Along this line of thought, we can simply count the frequency of each kind of amino acids in a protein sequence. However, this is not enough, because the amino acid "word/letter" could be composed of amino-acid pairs, three-amino-acid sequences, four-amino-acid sequences, and more-than-four-amino-acid sequences, so we need to count the frequency from amino-acid pairs to multi-amino-acid sequences.

As we do not know where an amino-acid "word/letter" begins and finishes, we may assume that an amino-acid "word/letter" could begin and finish anywhere in a protein sequence. Therefore we have to count the amino-acid sequence in the following way.

Along a protein sequence, any two amino acids in order can construct an amino-acid pair, i.e. the first and second; the second and third; the third and fourth, etc. Furthermore, any three amino acids in order can construct a three-amino-acid sequence, i.e. the first, second and third; the second, third and fourth; etc. The similar consideration can be applied to other multi-amino-acid sequences. This way, we can count the frequency of multi-amino-acid sequences. For instance, the human hemoglobin α-chain (accession number P01922) is composed of 142 amino acids, so it has 141 amino-acid pairs, 140 three-amino-acid sequences, 139 four-amino-acid sequences, and so on. Please note that each amino acid connects its neighboring amino acid in our counting, which is different from the concept of amino-acid pair in the secondary structure, where there is a distance between two amino acids.

2.2. Comparison of Counted Frequency with Reference

For analyzing an unknown language, we need to compare the frequency of possible unit (word) of the unknown language with that of a known one. Unfortunately, we do not have a known language regarding proteins. Thus, we need to establish a reference for comparison. The simple and reasonable way is to make a reference calculated according to the random principle, because pure chance is now considered to lie at the very heart of nature [17].

In an ideally random situation, two amino acids in an amino-acid pair can be constructed from any one of 20 kinds of amino acids, and there would be 400 (20^2) possible types of amino-acid pairs (Table 2-1). Similarly, if three amino acids in a three-amino-acid sequence could be randomly constructed from any one of 20 kinds of amino acids, there would be 8 000 (20^3) possible types of three-amino-acid sequences (Table 2-2). The similar deduction can be applied to other multi-amino-acid sequences, for example, there are 160 000 (20^4) possible types of four-amino-acid sequences, 3 200 000 (20^5) possible types of five-amino-acid sequences, and so on.

2.3. Counting Types of Amino-Acid Sequences in Real-Life Cases

As a protein is generally composed of 20 kinds of amino acids, we would expect to see each kind of amino acid in this protein. However, we do not know exactly whether or not a protein contains all 400 possible types of amino-acid pairs, 8 000 possible types of three-amino-acid sequences, and so on, although our intuition would tell us that a short protein sequence should not contain all of the possible types. For instance, the human hemoglobin α-chain should not contain all 400 possible types of amino-acid pairs, because it has only 141 amino-acid pairs.

Naturally, any amino-acid pair in a protein should be one of 400 possible types of amino-acid pairs, and any amino-acid pair, which does not appear in a protein, should also be one of 400 possible types of amino-acid pairs. Similarly, any three-amino-acid sequence in a protein should be one of 8 000 possible types of three-amino-acid sequences, and any three-amino-acid sequence, which does not appear in a protein, should also be one of 8 000 possible types of three-amino-acid sequences.

Table 2-1. 400 possible types of amino-acid pairs

	A	*R*	*N*	*D*	*C*	*E*	*Q*	*G*	*H*	*I*	*L*	*K*	*M*	*F*	*P*	*S*	*T*	*W*	*Y*	*V*
A	AA	AR	AN	AD	AC	AE	AQ	AG	AH	AI	AL	AK	AM	AF	AP	AS	AT	AW	AY	AV
R	RA	RR	RN	RD	RC	RE	RQ	RG	RH	RI	RL	RK	RM	RF	RP	RS	RT	RW	RY	RV
N	NA	NR	NN	ND	NC	NE	NQ	NG	NH	NI	NL	NK	NM	NF	NP	NS	NT	NW	NY	NV
D	DA	DR	DN	DD	DC	DE	DQ	DG	DH	DI	DL	DK	DM	DF	DP	DS	DT	DW	DY	DV
C	CA	CR	CN	CD	CC	CE	CQ	CG	CH	CI	CL	CK	CM	CF	CP	CS	CT	CW	CY	CV
E	EA	ER	EN	ED	EC	EE	EQ	EG	EH	EI	EL	EK	EM	EF	EP	ES	ET	EW	EY	EV
Q	QA	QR	QN	QD	QC	QE	QQ	QG	QH	QI	QL	QK	QM	QF	QP	QS	QT	QW	QY	QV
G	GA	GR	GN	GD	GC	GE	GQ	GG	GH	GI	GL	GK	GM	GF	GP	GS	GT	GW	GY	GV
H	HA	HR	HN	HD	HC	HE	HQ	HG	HH	HI	HL	HK	HM	HF	HP	HS	HT	HW	HY	HV
I	IA	IR	IN	ID	IC	IE	IQ	IG	IH	II	IL	IK	IM	IF	IP	IS	IT	IW	IY	IV
L	LA	LR	LN	LD	LC	LE	LQ	LG	LH	LI	LL	LK	LM	LF	LP	LS	LT	LW	LY	LV
K	KA	KR	KN	KD	KC	KE	KQ	KG	KH	KI	KL	KK	KM	KF	KP	KS	KT	KW	KY	KV
M	MA	MR	MN	MD	MC	ME	MQ	MG	MH	MI	ML	MK	MM	MF	MP	MS	MT	MW	MY	MV
F	FA	FR	FN	FD	FC	FE	FQ	FG	FH	FI	FL	FK	FM	FF	FP	FS	FT	FW	FY	FV
P	PA	PR	PN	PD	PC	PE	PQ	PG	PH	PI	PL	PK	PM	PF	PP	PS	PT	PW	PY	PV
S	SA	SR	SN	SD	SC	SE	SQ	SG	SH	SI	SL	SK	SM	SF	SP	SS	ST	SW	SY	SV
T	TA	TR	TN	TD	TC	TE	TQ	TG	TH	TI	TL	TK	TM	TF	TP	TS	TT	TW	TY	TV
W	WA	WR	WN	WD	WC	WE	WQ	WG	WH	WI	WL	WK	WM	WF	WP	WS	WT	WW	WY	WV
Y	YA	YR	YN	YD	YC	YE	YQ	YG	YH	YI	YL	YK	YM	YF	YP	YS	YT	YW	YY	YV
V	VA	VR	VN	VD	VC	VE	VQ	VG	VH	VI	VL	VK	VM	VF	VP	VS	VT	VW	VY	VV

A, alanine; R, arginine; N, asparagine; D, aspartic acid; C, cysteine; E, glutamic acid; Q, glutamine; G, glycine; H, histidine; I, isoleucine; L, leucine; K, lysine; M, methionine; F, phenylalanine; P, proline; S, serine; T, threonine; W, tryptophan; Y, tyrosine; V, valine.

Table 2-2. Mechanic table of 8 000 possible types of three-amino-acid sequences

AAA	AAR	...	AAI	AAL	...	AAY	AAV
ARA	ARR	...	ARI	ARL	...	ARY	ARV
...	...	...	...	...	...	...	...
IVA	IVR	...	IVI	IVL	...	IVY	IVV
LAA	LAR	...	LAI	LAL	...	LAY	LAV
...	...	...	...	...	...	...	...
VYA	VYR	...	VYI	VYL	...	VYY	VYV
VVA	VVR	...	VVI	VVL	...	VVY	VVV

Tables 2-3 list several real-life examples on counting types of amino-acid sequences in proteins with different lengths. First of all, we can see that the number of counted types increases as the protein becomes longer and longer (compare each column in the table). Additionally, we can see that the number of counted types is different from the number that a protein has, that is, each type of amino-acid sequence can repeat in a protein. For example, human hemoglobin α-chain has 142 amino acids, 141 amino-acid pairs, 140 three-amino-acid sequences, 139 four-amino-acid sequences, but in reality it contains 100 types of amino-acid pair, 138 types of three-amino-acid sequence, 137 types of four-amino-acid sequence, because some types appear more than once.

Table 2-3. Total amino acids and number of types of amino-acid pairs, three-amino-acid sequences and four-amino-acid sequences in some proteins

Protein	Total	II	III	IV	Ref
Human hemoglobin α-chain (P01922)	142	100	138	139	[22]
Human tumor necrosis factor (P01375)	233	157	226	230	[23]
Mouse p53 protein (P02340)	390	212	370	387	[24]
Human acute myeloid leukemia 1 protein (Q01196)	453	218	424	448	[25]
Human tyrosinase (P14679)	529	286	505	524	[26]
Human dopamine β-hydroxylase (P09172)	603	283	565	597	[27]
HIV envelope polyprotein GP160 (P03378)	855	314	783	848	
Human Na$^+$/K$^+$ ATPase α-1 chain (P05023)	1023	316	913	1008	
Human insulin receptor (P06213)	1382	355	1192	1364	

II, amino-acid pairs, III, three-amino-acid sequences, IV, four-amino-acid sequences, Ref, references.

2.4. Counting Amino-Acid Pairs with Respect to 400 Possible Types

From the above example, we can see that a protein generally does not contain all possible types of multi-amino-acid sequences, although the longer a protein is, the more types a protein contains.

After counting the types of multi-amino-acid sequences of a protein in question, we have at least the concept of present and absent types in the protein, which suggests which types a protein favors and which types a protein does not favor. Functionally, we may assume that the present types are necessary for the protein survival, whereas the absent types are somewhat unnecessary or harmful.

Our counting is somewhat similar to the determination of present and absent word/letter in a sentence. Naturally, the next step is to determine the frequency of each letter because we know that some letters appear more frequently. Still, Table 2-3 indicates the repetition of multi-amino-acid sequences, which is more observable in types of amino-acid pairs.

In total, thirty-one types of amino-acid pairs appear more than once in human hemoglobin α-chain (Table 2-4). In addition, 2 types of three-amino-acid sequences "ALS" and "LSH" appear twice. However, there is no repetition in the types of more-than-three-amino-acid sequences in human hemoglobin α-chain.

It is very suggestive that the types of amino-acid pairs repeat far more frequently than those of multi-amino-acid sequences, which appear so rarely that we might not need to spend great energy to analyze them. This feature may imply that the more-than-two-amino-acid sequence has a small chance of being "word/letter". Therefore, we can concentrate our analysis on the amino-acid pairs, which is plausible because a good signature pattern of a protein must be as short as possible and many short sequences (not more than four or five residues long conserved sequence) are often diagnostics of certain finding properties or active sites [28].

Table 2-4. Repetition of types of amino-acid pairs in human hemoglobin α-chain

Type	Repetition	Type	Repetition	Type	Repetition	Type	Repetition
AA	2	DL	2	LS	6	SA	2
AD	2	DK	2	LT	2	SH	2
AE	2	GA	2	KL	2	TN	2
AH	4	GK	2	KT	2	VA	2
AL	4	HA	3	KV	2	VD	2
AS	2	HG	2	FL	2	VL	2
AV	2	LA	2	FP	2	VK	2
NA	2	LL	2	PA	3		

Still, although a point mutation is related to a single amino acid, both original and mutated amino acids (except for the one at terminal) have connections with two neighboring amino acids, which constructs two amino-acid pairs, thus the use of amino-acid pair takes into account the interaction between neighboring amino acids.

2.5. Numerical Comparison with Reference

Now we have already established the reference for comparison, that is, we use the counted types of amino-acid pairs in a protein to compare with 400 possible types of amino-acid pairs, by which we know the present and absent types of amino-acid pairs in a protein.

For the development of a method, we should always look at if there is the possibility of improving the current version of methods. In this sense, we can see that the comparison with 400 possible types of amino-acid pairs is a semi-quantitative measure, because the same type of amino-acid pair can repeat several times (Table 2-4).

Therefore we need to consider not only whether a type of amino-acid pair appears in a protein, but also how many times it appears. By these considerations, we can have the following calculations, which give the attribute to the amino-acid pair predictability. We still use the human hemoglobin α-chain as the example because it is relatively short and important. Bear in mind, the human hemoglobin α-chain is composed of 142 amino acids (Table 2-5), which construct 141 amino-acid pairs.

Table 2-5. Amino-acid composition of human hemoglobin α-chain

Amino acid	Number
Alanine, A	21
Arginine, R	3
Asparagine, N	4
Aspartic acid, D	8
Cysteine, C	1
Glutamic Acid, E	4
Glutamine, Q	1
Glycine, G	7
Histidine, H	10
Isoleucine, I	0
Leucine, L	18
Lysine, K	11
Methionine, M	3
Phenylalanine, F	7
Proline, P	7
Serine, S	11
Threonine, T	9
Tryptophan, W	1
Tyrosine, Y	3
Valine, V	13

2.5.1. Attribute I of Amino-Acid Pair Predictability

There are 21 alanines "A" and 11 serines "S" in human hemoglobin α-chain. Let us consider how to make the prediction of amino-acid pair according to the permutation. Imaginably, we have a bag containing 142 balls, of which each is marked with the letters of amino acids, so we have 21 balls marked as "A", 11 balls marked as "S" and so on. These 142 balls are unordered mixed together in a bag. We randomly pick a ball from the bag, the chance of picking a ball marked with "A" is 21/142, and then we randomly pick the second ball from the bag, the chance of picking a ball marked with "S" is 11/141 because this time there are 141 balls left in the bag.

Now we have two unrelated events, whose combined probability is the product of probabilities of both events, so the probability of picking the ball marked with "A" and picking the ball marked with "S" is 21/142×11/141=0.0115. This means that there is 0.0115 chance of finding "AS" among 141 amino-acid pairs in human hemoglobin α-chain, which is equal to 0.0115×141=1.627.

Taking together, we can predict the frequency of amino-acid pairs in human hemoglobin α-chain through such a way. The frequency of random presence of amino-acid pair "AS" is 2 after rounding up to integral (21/142×11/141×141=1.627), that is, the "AS" would appear twice in the human hemoglobin α-chain. Actually, we do find two "AS" in this chain, one locates at positions 124 and 125 and the other at positions 131 and 132, so the actual frequency of "AS" is 2. In this case, we have predicted the presence of the type "AS" according to the permutation, and we have also predicted the frequency of "AS", thus both the type of amino-acid pair and its frequency are predictable, and the difference between its actual and predicted frequency should be zero.

After above process and calculation, we put these two balls into the bag again, and make another similar process. Here we have two types of probabilistic approaches, (i) returning two balls into the bag, and each guess begins with all balls in the bag, and (ii) not returning two balls into the bag, and each guess begins with less and less balls in the bag. Nevertheless, the second approach is probabilistically far more complicated. Thus, we currently use the first approach, because our goal is not the mechanism of randomness, but to use a value to present a living protein.

We can call this calculated case as the randomly predictable present type of amino-acid pair with predictable frequency, or simply as attribute I of amino-acid pair predictability.

2.5.2. Attribute II of Amino-Acid Pair Predictability

There are ten histidines "H" and one glutamine "Q" in human hemoglobin α-chain. The predicted frequency of "HQ" is 0 (10/142×1/141×141=0.0704), i.e. the "HQ" would not appear in this human hemoglobin α-chain, which is true in the real situation. Thus the absence of the type "HQ" is predictable, and the difference between its actual and predicted frequency is 0.

We call this calculated case as the randomly predictable absent type of amino-acid pair, or simply as attribute II of amino-acid pair predictability.

2.5.3. Attribute III of Amino-Acid Pair Predictability

Now let us look at another amino-acid pair in human hemoglobin α-chain, as we already know that there are 21 alanines "A" in human hemoglobin α-chain. According to the process above, the frequency of random presence of amino-acid pair "AA" is 3 (21/142×20/141×141=2.9577), i.e. there would be three "AA" in human hemoglobin α-chain. Actually, the "AA" appears twice at positions 13 and 14 and at positions 111 and 112, so the presence of the type "AA" is predictable, but its frequency is unpredictable, and the difference between its actual and predicted frequency is –1.

This is the case that the actual frequency is smaller than its predicted one. Another case is that the actual frequency is larger than its predicted one, for instance, there are 8 aspartic acids "D" in the human hemoglobin α-chain, the predicted frequency of "AD" is 1 ($21/142 \times 8/141 \times 141 = 1.1831$), while its actual frequency is 2 at positions 6 and 7 and at positions 64 and 65, so the difference between its actual and predicted frequency is 1.

We call these two calculated cases as the randomly predictable present type of amino-acid pair with unpredictable frequency, or simply as attribute III of amino-acid pair predictability.

2.5.4. Attribute IV of Amino-Acid Pair Predictability

There are 10 histidines "H" and 7 glycines "G" in human hemoglobin α-chain, and the predicted frequency of "HG" is 0 ($10/142 \times 7/141 \times 141 = 0.4930$), so the type "HG" would not appear in this chain. However, it appears twice in the reality at positions 51 and 52 and at positions 59 and 60, thus the presence of the type "HG" is unpredictable. Naturally, its frequency is unpredictable too, and the difference between its actual and predicted frequency is 2.

We call this calculated case as the randomly unpredictable present type of amino-acid pair, or simply as attribute IV of amino-acid pair predictability.

2.5.5. Attribute V of Amino-Acid Pair Predictability

There are 21 alanines "A" and 11 lysines "K" in human hemoglobin α-chain. The predicted frequency of "AK" is 2 ($21/142 \times 11/141 \times 141 = 1.6268$), i.e. there would be two "AK" in the human hemoglobin α-chain. However, no "AK" appears in the reality, therefore the absence of the type "AK" from the human hemoglobin α-chain is unpredictable. Naturally, its frequency is unpredictable too, and the difference between its actual and predicted frequency is –2.

We call this calculated case as the randomly unpredictable absent type of amino-acid pair, or simply as attribute V of amino-acid pair predictability.

2.5.6. Predictable and Unpredictable Portions of Amino-Acid Pairs

After the calculations described above, the amino-acid pairs in a protein can be classified as predictable and unpredictable portions with respect to the type as well as to the frequency, and the sum of both predictable and unpredictable portions is 100%. The predictable portion in type refers to (i) how many predictable types a protein have with respect to the 400 possible types of amino-acid pairs, and (ii) how many predictable types a protein have with respect to the present types. The predictable portion in frequency refers to how many pairs of present types a protein has. Of the amino-acid pairs in human hemoglobin α-chain, the predictable and unpredictable portions are 74.75% and 25.25% in 400 possible types, 38% and 62% in present types, as well as 31.21% and 68.79% in frequencies, respectively. It should be noted that the absent types of amino-acid pair are included in 400 possible types and that a present type of amino-acid pair can appear several times in a protein. Both predictable and unpredictable portions can serve as a quantitative measure to present a whole protein.

This way, we generally have two types of measures for comparison: (1) we can use the predicted or unpredicted portion if we would like to compare the difference among whole proteins, say, each protein serves as a unit for comparison, and (2) we can use the difference between actual and predicted frequency for a type of amino-acid pair if we would like to compare the difference cross different types of amino-acid pairs, say, each amino-acid pair serves as a unit for comparison.

2.6. Visualization of Amino-Acid Pair Predictability

With the advance of computation, we can have not only the computed data but also their graphic presentation. Herein we visualize the amino-acid pair predictability to see what it is.

Taking human hemoglobin α-chain as an example, we have counted the actual frequency of 141 amino-acid pairs in this chain, and then calculated the predicted frequency of amino-acid pairs based on the amino-acid composition. Actually, we have already defined the positions of present amino-acid pairs in the above calculations.

The advantage of our methods is to deal with the primary structure of a protein, thus we only need a 2-dementional graphic presentation, which is much easier than the presentation of 3-dementional graphics. Thus, we can simply make a 2-dementional graph, where the x-axis is the position of the protein in question, in our case, the x-axis ranges from 0 to 143 because the human hemoglobin α-chain has 142 amino acids, while the y-axis presents what we are interested, for example, the actual frequency of amino-acid pairs (the upper panel in Figure 2-1), the predicted frequency of amino-acid pairs (the middle panel in Figure 2-1), and the difference between actual and predicted frequency (the lower panel in Figure 2-1).

In fact, either the actual frequency or predicted one only presents one of two pieces of information from different angle. On the other hand, the difference between actual and predicted frequency represents the combination of information provided by both the actual and predicted frequencies. Therefore, we prefer to use the difference between actual and predicted frequency of amino-acid pairs (the lower panel in Figure 2-1) to visualize the computed data.

Generally, we can use the division to combine two pieces of information together, for example, (predicted frequency)/(actual frequency), however, we prefer to use reduction, i.e. (actual frequency)–(predicted frequency), because it is often that either actual frequency or predicted frequency is zero, which leads the difficulty in division.

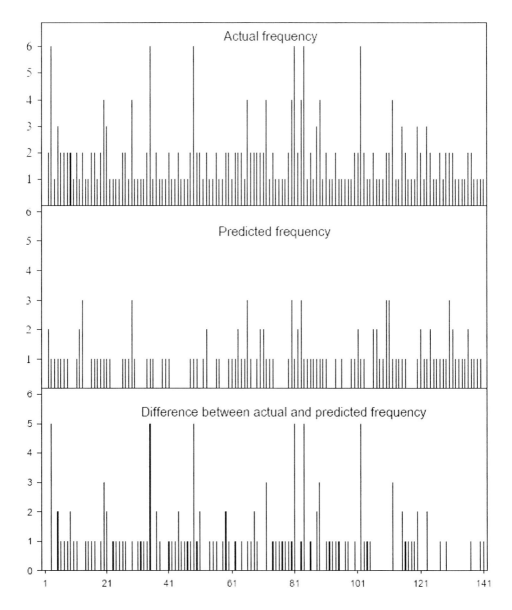

Figure 2-1. Visualization of amino-acid pair predictability in human hemoglobin α-chain.

Generally, we can use the division to combine two pieces of information together, for example, (predicted frequency)/(actual frequency), however, we prefer to use reduction, i.e. (actual frequency)–(predicted frequency), because it is often that either actual frequency or predicted frequency is zero, which leads the difficulty in division.

Another advantage of using the difference between actual and predicted frequency is that this visualization assigns the chance of occurrence of mutation along a protein, as we will see later that the randomness can be considered as a power engineering mutation. Frankly speaking, there is no general rule on how to combine two pieces of information together. It depends on researchers themselves and how the combination can facilitate the research.

We are used to assign the difference between actual and predicted frequency at the second amino acid of an amino-acid pair, and thus the first amino acid in a protein is generally not assigned, but we generally assign this position with unity. Still our experience shows that the calculation becomes easier when a protein with the length around 400.

Chapter 3

Behavior of Amino-Acid Pair Predictability

So far we have developed a method to measure the randomness with respect to amino-acid pair in a protein. In order to determine whether this measure is useful, we need to observe the behavior of amino-acid pair predictability in various situations. Here, the behavior has two practical meanings, (i) we need to observe whether the amino-acid pair predictability is different cross proteins, cross subtypes, cross species, etc., and (ii) we need to observe whether the amino-acid pair predictability is different in a protein along the time course, such as before and after mutation. Had we got the answer that the amino-acid pair predictability does not differ among various situations, we would conclude that this measure is constant. If not so, we can conclude that the measure we developed is alive and dynamic.

3.1. Behavior of Amino-Acid Pair Predictability from Spatial Angle

The differences in functions and evolutionary speeds lead the proteins to be different one another. Thus the question here is whether the amino-acid pair predictability can demonstrate these differences, or in other words whether the amino-acid pair predictability is different from a protein to another one. To test this hypothesis, we can calculate several proteins from different sources and compare them.

3.1.1. Amino-Acid Pair Predictability in Eight Proteins

Now we look at several amino-acid pairs in eight different proteins, which are not specially selected for this purpose, but were analyzed in our previous studies [29-36]. Table 3-1 shows the amino-acid composition of these eight proteins. As seen, their compositions are different one from another.

According to Table 3-1, we can count the actual frequency of any type of amino-acid pairs of interests, calculate their predicted frequency, and determine their attributes.

Table 3-1. Amino-acid compositions in eight proteins

Protein	AT7B	CA54	FA9	GLCM	HBA	LDLR	PH4H	VHL
Accession number	P35670	P29400	P00740	P04062	P01922	P01130	P00439	P40337
Reference	[29]	[30]	[31]	[32]	[33]	[34]	[35]	[36]
A	134	47	23	42	21	43	28	10
R	53	37	20	23	3	46	24	20
N	49	32	32	19	4	41	16	9
D	62	54	19	26	8	75	24	11
C	29	20	24	8	1	63	9	2
E	80	61	43	21	4	48	36	30
Q	69	73	14	21	1	34	20	8
G	104	478	36	37	7	62	25	18
H	36	13	10	19	10	19	13	5
I	106	69	25	22	0	36	28	6
L	132	112	28	60	18	66	50	20
K	75	80	28	23	11	36	30	3
M	44	26	6	11	3	11	3	3
F	36	41	21	27	7	26	27	5
P	72	391	15	37	7	39	23	19
S	126	64	27	44	11	70	28	11
T	82	38	30	32	9	50	24	7
W	11	5	7	13	1	20	3	3
Y	24	13	16	19	3	17	22	6
V	141	31	37	32	13	58	19	17
Total	1465	1685	461	536	142	860	452	213

A, alanine; R, arginine; N, asparagine; D, aspartic acid; C, cysteine; E, glutamic acid; Q, glutamine; G, glycine; H, histidine; I, isoleucine; L, leucine; K, lysine; M, methionine; F, phenylalanine; P, proline; S, serine; T, threonine; W, tryptophan; Y, tyrosine; V, valine.

AT7B, human copper-transporting ATPase 2; CA54, human collagen α5(IV) chain precursor; FA9, human coagulation factor IX precursor; GLCM, human glucosylceramidase precursor; HBA, human hemoglobin α-chain; LDLR, human low-density lipoprotein receptor precursor; PH4H, human phenylalanine-4-hydroxylase; VHL, human Von Hippel-Lindau disease tumor suppressor.

In Table 3-2, we calculate the predicted frequency of amino-acid pair "AA", count its actual frequency and determine its attribute in these eight different proteins. Although "AA" appears in seven proteins, its predicted and actual frequencies are very different. Or we can say that the predicted and measured randomness is different from protein to protein although "AA" in nature is not different either in its physical or chemical property in different proteins.

Because the amino-acid pair "AA" in Table 3-2 does not include all of attributes, we look at another amino-acid pair "MM" in Table 3-3. Although "MM" does not appear in seven proteins, its predicted frequency is indeed different one from another.

These three tables show that the behavior of amino-acid pair predictability does vary among these eight proteins.

Table 3-2. Calculation of amino-acid pair "AA" in eight different proteins

Protein	Total amino acids	Number of "A"	Predicted frequency of "AA"	Actual frequency	Attribute
AT7B	1465	134	134/1465×133/1464×1464=12.1652≈12	12	I
CA54	1685	47	47/1685×46/1684×1684=1.2831≈1	4	III
FA9	461	23	23/461×22/460×460=1.0976≈1	2	III
GLCM	536	42	42/536×41/535×535=3.2127≈3	3	I
HBA	142	21	21/142×20/141×141=2.9577≈3	2	III
LDLR	860	43	43/860×42/859×859=2.100≈2	3	III
PH4H	452	28	28/452×27/451×451=1.6726≈2	1	III
VHL	213	10	10/213×9/212×212=0.4225≈0	0	II

I, randomly predictable present type of amino-acid pair with predictable frequency; II, randomly predictable absent type of amino-acid pair; III, randomly predictable present type of amino-acid pair with unpredictable frequency.

Table 3-3. Calculation of amino-acid pair "MM" in eight different proteins

Protein	Total amino acids	Number of "M"	Predicted frequency of "MM"	Actual frequency	Attribute
AT7B	1465	44	44/1465×43/1464×1464=1.2912≈1	0	V
CA54	1685	26	26/1685×25/1684×1684=0.3858≈0	2	IV
FA9	461	6	6/461×5/460×460=0.0651≈0	0	II
GLCM	536	11	11/536×10/535×535=0.2052≈0	0	II
HBA	142	3	3/142×2/141×141=0.0423≈0	0	II
LDLR	860	11	11/860×10/859×859=0.1279≈0	0	II
PH4H	452	3	3/452×2/451×451=0.0133≈0	0	II
VHL	213	3	3/213×2/212×212=0.0282≈0	0	II

II, randomly predictable absent type of amino-acid pair; IV, randomly unpredictable present type of amino-acid pair; and V, randomly unpredictable absent type of amino-acid pair.

3.1.2. Predictable Portion in Proteins with Different Lengths

In the above section, we have seen the behavior of amino-acid pair predictability in eight different proteins. Actually, the actual and predicted frequencies are more suitable for observing the behavior of amino-acid pair predictability in the amino-acid pairs of interests. On the other hand, we have the predictable and unpredictable portions of amino-acid pairs to describe the whole protein. Thus, we can use them if we are interested in the comparison of whole proteins.

As the amino-acid pair predictability is different among various proteins, we would expect to the predictable and unpredictable portions different in different proteins. However, we must check this expectation against the reality.

In the above section, the eight proteins have different lengths, thus the first task we are facing is to determine whether the predictable portion is different among various proteins with different lengths. The dataset we will use contains 74 proteins from different species with different functions and lengths (Table 3-4).

Table 3-4. Dataset used for calculation of predictable portion of amino-acid pairs

Length	Protein type	Accession no.	Source	Species	References
76	Envelope protein	P59637	Corona virus	Human	[37,38]
82	Envelope protein	P24415	Corona virus	Porcine	[37]
84	Nonstructural protein	Q04854	Corona virus	Human	[37]
97	Matrix protein 2	Q9Q0L9	Influenza A virus	Goose	[39]
97	Matrix protein 2	Q9EAF2	Influenza A virus	Human	[39]
101	Nonstructural protein	Q04703	Corona virus	Canine	[37]
109	Nonstructural protein	Q04853	Corona virus	Human	
121	Nonstructural protein 2	Q9Q0L7	Influenza A virus	Goose	[39]
121	Nonstructural protein 2	Q9DHF8	Influenza A virus	Human	[39]
142	Hemoglobin α-chain	P01922	Human	Human	[33, 40]
176	Nonstructural protein	P33467	Corona virus	Feline	[37]
213	Von Hippel-Lindau disease tumor suppressor	P40337	Human	Human	[36]
213	Nonstructural protein	Q04704	Corona virus	Canine	[37]
221	Membrane protein	P59596	Corona virus	Human	[38]
228	Membrane protein	P03415	Corona virus	Murine	[37]
230	Nonstructural protein 1	Q9Q0L6	Influenza A virus	Goose	[39]
230	Nonstructural protein 1	Q9DHF7	Influenza A virus	Human	[39]
252	Matrix protein 1	Q9Q0L8	Influenza A virus	Goose	[39]
252	Matrix protein 1	Q91U69	Influenza A virus	Human	[39]
277	Nonstructural protein	P18517	Corona virus	Bovine	[37]
307	Replicase polyprotein 1ab	Q9WQ77	Corona virus	Rat	[37]
309	Pilx9 protein	Q9EUF2	*Escherichia coli*	*Escherichia coli*	
318	Andr	AY223539	*Burkholderia cepacia*	*Burkholderia cepacia*	
354	Ornithine carbamoyltransferase	P00481	Rat	Rat	
354	Ornithine carbamoyltransferase	P00480	Human	Human	
354	Ornithine carbamoyltransferase	P11725	Mouse	Mouse	
376	Nucleocapsid protein	O12298	Corona virus	Feline	[37]
382	Nucleocapsid protein	Q04700	Corona virus	Feline	[37]
393	P53	P04637	Human	Human	[41]
405	Andr	NC_007435	*Burkholderia pseudomallei 1710b*	*Burkholderia pseudomallei 1710b*	
422	Nucleocapsid protein	P59595	Corona virus	Human	[38]
423	Hemagglutinin-esterase precursor	AY316300	Corona virus	Equine	
424	Hemagglutinin-esterase precursor	P30215	Corona virus	Human	[37]
448	Nucleocapsid protein	P10527	Corona virus	Bovine	[37]

Length	Protein type	Accession no.	Source	Species	References
452	Phenylalanine-4-hydroxylase	P00439	Human	Human	[35]
461	Coagulation factor IX precursor	P00740	Human	Human	[31]
467	Neuraminidase	Q9ICY2	Influenza A virus	Human	[39]
469	Neuraminidase	Q9Q0U7	Influenza A virus	Goose	[39]
498	Nucleoprotein	Q9Q0U8	Influenza A virus	Goose	[39]
498	Nucleoprotein	Q9ICY7	Influenza A virus	Human	[39]
536	Glucosylceramidase precursor	P04062	Human	Human	[32]
560	Hemagglutinin	Q9ICY5	Influenza A virus	Human	[39, 42]
568	Hemagglutinin	Q9Q0U6	Influenza A virus	Goose	[39, 42]
659	Bruton's tyrosine kinase	Q06187	Human	Human	[43]
659	Bruton's tyrosine kinase	P35991	Mouse	Mouse	
716	Polymerase acidic protein	Q9Q0U9	Influenza A virus	Goose	[39, 44]
716	Polymerase acidic protein	Q9ICX4	Influenza A virus	Human	[39, 44]
757	Polymerase basic protein 1	Q9Q0V0	Influenza A virus	Goose	[39, 44]
758	Polymerase basic protein 1	Q9ICX7	Influenza A virus	Human	[39, 44]
759	Polymerase basic protein 2	Q9Q0V1	Influenza A virus	Goose	[39, 44]
759	Polymerase basic protein 2	Q9ICY1	Influenza A virus	Human	[39, 44]
860	Low-density lipoprotein receptor precursor	P01130	Human	Human	[34]
864	Low-density lipoprotein receptor precursor	P35951	Mouse	Mouse	
879	Low-density lipoprotein receptor precursor	P35952	Rat	Rat	
899	Androgen receptor	P19091	Mouse	Mouse	
899	Androgen receptor	NP_038504	Mouse	Mouse	
902	Androgen receptor	P15207	Rat	Rat	
917	Androgen receptor	AAA51772	Human	Human	
919	Androgen receptor	P10275	Human	Human	
972	Structural polyprotein	Q703G9	Infectious pancreatic necrosis virus – Sp	Infectious pancreatic necrosis virus – Sp	
1114	Proto-oncogene tyrosine-protein kinase receptor ret precursor	P07949	Human	Human	
1130	Proto-oncogene tyrosine-protein kinase	AAB60394	Human	Human	

Table 3-4. Continued

Length	Protein type	Accession no.	Source	Species	References
1173	Spike glycoprotein precursor	P15423	Corona virus	Human	[38]
1208	Spike glycoprotein precursor	AY342357	Corona virus	Turkey	
1255	Spike glycoprotein precursor	P59594	Corona virus	Human	[38, 45]
1360	Spike glycoprotein precursor	Q9IKD1	Corona virus	Rat	[37]
1379	Met proto-oncogene tyrosine kinase	P16056	Mouse	Mouse	
1382	Met proto-oncogene tyrosine kinase	P97523	Rat	Rat	
1384	Met proto-oncogene tyrosine kinase	Q769I5	Cattle	Cattle	
1390	Met proto-oncogene tyrosine kinase	P08581	Human	Human	
1464	Collagen α1(I) chain precursor	P02452	Human	Human	[46]
1465	Copper-transporting ATPase 2	P35670	Human	Human	[29]
1466	Collagen α1(III) chain precursor	P02461	Human	Human	[47]
1685	Collagen α5(IV) chain precursor	P29400	Human	Human	[30]

The calculated predictable portions for these 74 proteins are presented in Figure 3-1, which shows three types of predictable portions.

1. The black cycles are the predictable portion with respect to 400 possible types of amino-acid pairs. Actually, each black cycle means how many amino-acid pairs, whose actual frequency is equal to its predicted frequency (AF=PF) among 400 possible types of amino-acid pairs. In this sense, the black cycles include both randomly predictable present types of amino-acid pairs with predictable frequency (Attribute I) and randomly predictable absent types of amino-acid pairs (Attribute II). As seen in this figure, the predictable portion decreases as the protein length increases from 76 to around 400. Without exploring the underlined reason, the predictable portion is indeed different one another.

2. The gray squares are the predictable portion with respect to the type of amino-acid pairs, which can really be found in the protein. In this sense, they include only the randomly predictable present types of amino-acid pair with predictable frequency (Attribute I). Thus, each gray square means how many amino-acid pairs, whose actual frequency is equal to its predicted one (AF=PF>0) among present types of amino-acid pairs in a protein. As seen in this figure, the predictable portion is

different for these 74 proteins, and a slight pattern can be seen as the protein length increases.

3. The hollow triangles are the predictable portion with respect to the frequency of amino-acid pairs, which can also be found in the protein. In this sense, this counts all amino-acid pairs belonging to the randomly predictable present types of amino-acid pair with predictable frequency (Attribute I). As can be seen in this figure, the hollow triangles have a similar pattern as the gray squares. The difference between points 2 and 3 is that we count the types in point 2 but the pairs in point 3, and both the types and pairs are predictable.

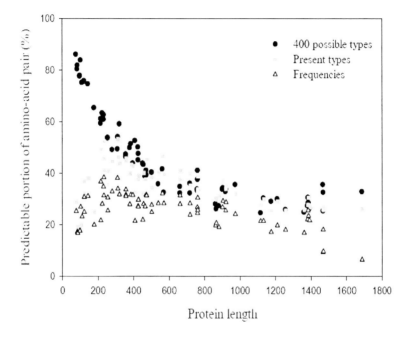

Figure 3-1. Predictable portions of 74 proteins with different length from different species.

Taking together, we can conclude that the difference between the black cycles and others is due to the fact that the black cycles include both present (Attribute I, AF=PF>0) and absent types (Attribute II, AF=PF=0), thus they reveal a different pattern from the other two groups, which include only the present types or frequencies (Attribute I, AF=PF>0).

In Sections 3.2.4 and 3.2.6, we will have more detailed discussion of how to calculate the predictable portions. For this section, we have observed the behavior of predictable portion among 74 proteins with different lengths.

3.1.3. Predictable Portion in Proteins with Similar Length

In this section, we make further observation on the behavior of predictable portion of amino-acid pairs because science requires us to validate a concept in as many scenarios as possible on the one hand, and research requires us to apply a concept to as many topics as possible on the other hand.

Now we use a huge database to observe the behavior of predictable portion, because our recent studies concentrate on proteins from influenza A virus, whose database is increasing significantly due to the fact that possible epidemic/pandemic of influenza in human [48-56]. By accumulating point mutations (genetic drift), the influenza A viruses change their antigenic properties mainly in the RNA genes coding for the surface proteins, hemagglutinin and neuraminidase [57].

The proteins we use are the hemagglutinins from influenza A virus, which have the same function and include sixteen subtypes from H1 to H16 (Table 3-5). We can see that the protein length for a subtype is similar, as the coefficient of variation (CV) is quite small. In particular, the length is identical for H10, H12 and H15, respectively. Thus these data can be considered to meet our request for observing the behavior of predictable portion.

Table 3-5. Average length of 2495 hemagglutinins from influenza A virus

Hemagglutinin subtype	Number of protein	Mean of length	Standard deviation (SD)	Coefficient of variation (%)
H1	408	565.53	0.52	0.09
H2	46	562.76	2.18	0.39
H3	983	565.97	0.25	0.04
H4	23	563.00	2.84	0.51
H5	521	567.36	1.59	0.28
H6	61	565.93	0.60	0.11
H7	152	561.30	4.20	0.75
H8	10	565.90	0.32	0.06
H9	163	559.93	0.59	0.11
H10	28	561.00	0.00	0.00
H11	62	564.98	0.13	0.02
H12	17	564.00	0.00	0.00
H13	6	565.50	0.55	0.10
H14	2	565.50	3.54	0.63
H15	7	570.00	0.00	0.00
H16	6	564.83	0.75	0.13

Figure 3-2 displays the predictable portions of amino-acid pairs with respect to 400 possible types, present types and frequencies. This figure can be read as follows. The x- and y-axes indicate the hemagglutinin length and predictable portion in different contexts. A filled cycle marks the mean of length in given hemagglutinins with reference to x-axis, and the mean of predictable portion with reference to y-axis. Similarly, a capped line along x-axis direction is the standard deviation (SD) of length in given hemagglutinins, and a capped line along y-axis direction is the standard deviation (SD) of predictable portion. As we use the same means and SDs for x-axis for each panel, we only mark the hemagglutinin subtypes in the upper panel.

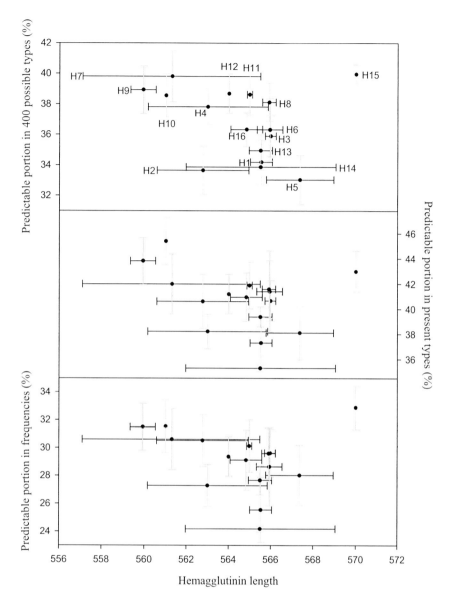

Figure 3-2. Predictable portions of amino-acid pairs versus sequence length in sixteen subtypes of hemagglutinins from influenza A virus. The data are presented as mean±SD.

For example, we can find H5 hemagglutinins in the lowest position in the upper panel, whose mean of length is 566.51 and SD is 4.45, which are obtained by calculating 521 H5 hemagglutinins (Table 3-5). If we draw a line along position 566.51 from the upper panel down through the middle and lower panels, we can see two filled cycles that are H5 hemagglutinins, too. Thus for H5 hemagglutinins, the predictable portions of amino-acid pairs are 33.09±1.71 (mean ± SD) in 400 possible types (the upper panel), 38.15±2.18 in present types (the middle panel), and 28.07±2.19 in frequencies (the lower panel). In other words, Figure 3-5 shows that less than 3% change in hemagglutinin length would lead to about 10% variations in the predictable portion of amino-acid pairs.

Actually, Figure 3-2 reveals the behavior of predictable portion different in the proteins with similar length, even in those with the same function.

3.1.4. Predictable Portion in Various Subtypes of Proteins

With the progress of our observation, we now check the behavior of predictable portion of amino-acid pairs in different subtypes of proteins. Actually, we have already dealt partially with this issue in the last section, where we focus on the similar length of proteins.

In Figure 3-3, we show the predictable portions in different subtypes of hemagglutinins from influenza A virus. In fact, this type of figure is more frequently used in research front for comparison. As seen, the predictable portion is different from a subtype to another. We can find out the difference among subtypes significant ($p<0.001$) when we use the one-way ANOVA (analysis of variance) to test the statistical difference.

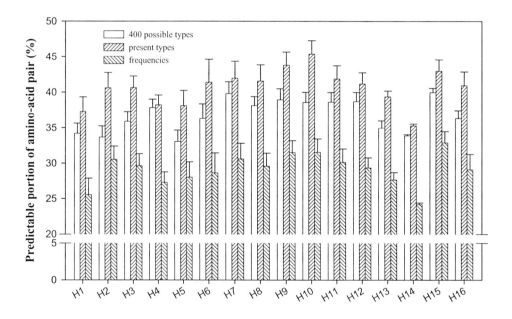

Figure 3-3. Predictable portions of amino-acid pairs in different subtypes of hemagglutinins from influenza A virus (the number of hemagglutinins in each subtypes can be found in Table 3-5). The data are presented as mean±SD.

Over recent years, the H5N1 influenza A viruses have drawn a considerable attention in public domain [58-62]. In Figure 3-3, we can see some difference between H5 subtype and others. In general, the predictable portion in H5 subtype is smaller than others if we consider the sample size of H5 hemagglutinins is very large (Table 3-5). A small predictable portion suggests the protein construction is less randomly or less parsimony, thus nature would have the trend of making H5 hemagglutinins more economic, more predictable, through mutations. This is one of the reasons for frequent mutations found in H5 hemagglutinins from our random viewpoint [39, 42, 63, 64].

3.1.5. Predictable Portion in Proteins cross Species

Finally we look at the behavior of predicable portion of amino-acid pairs in proteins cross species to finish our observation from spatial angle. We still use the dataset of hemagglutinins from influenza A virus in above sections. These data can be grouped according to their species (Table 3-6).

**Table 3-6. 2495 unique hemagglutinins from influenza
|A virus grouped according to species**

Subtype	Species	Year of isolation	Number
H1	Avian	1976-2002	42
	Human	1918-2006	295
	Swine	1930-2004	71
H2	Avian	1961-1998	19
	Human	1957-2005	27
H3	Avian	1969-2005	33
	Equine	1963-1994	37
	Human	1968-2006	886
	Seal	1992	2
	Swine	1971-2005	25
H4	Avian	1956-2000	20
	Seal	1982	1
	Swine	1999	2
H5	Avian	1959-2006	451
	Cat	2004-2006	2
	Human	1997-2006	54
	Leopard	2004	1
	Swine	2001-2004	7
	Tiger	2004	3
	Unknown	2001	3
H6	Avian	1972-2004	61
H7	Avian	1927-2006	137
	Equine	1956-1977	12
	Human	1996	1
	Seal	1980	2
H8	Avian	1968-2005	10

Table 3-6. Continued

Subtype	Species	Year of isolation	Number
H9	Avian	1966-2004	159
	Human	1999	2
	Swine	2003-2004	2
H10	Avian	1949-2006	27
	Mink	1984	1
H11	Avian	1956-2006	62
H12	Avian	1976-2005	17
H13	Avian	1977-1984	5
	Whale	1984	1
H14	Avian	1982	2
H15	Avian	1979-1983	7
H16	Avian	1975-1999	6

With these data, we can illustrate the predictable portions cross species in Figure 3-4, and several features can be found in this figure. First, the predictable portions of amino-acid pairs vary significantly from one species to another, except for the H4 and H5 hemagglutinins. Second, equine hemagglutinins have generally the largest predictable portion among different species, implying that their construction is more randomly. Third, human hemagglutinins have smaller predictable portion than others especially in H1 and H2 subtypes, indicating that the construction of human hemagglutinins is less randomly or less parsimony.

By the way, we may slightly relate the results in Figure 3-4 to the hot research topic, influenza. As seen in this figure, the difference between avian and human is the least in H5 subtype, which indicates the easy transmission of influenza A virus from avian to human. Our analysis can throw light on understanding why many human cases were directly affected by avian H5N1 virus [65-67].

Hence we can say the behavior of predictable portion of amino-acid pairs different cross species.

3.2. Behavior of Amino-Acid Pair Predictability from Time Angle

The evolutionary process is going along the time course, therefore the behavior of amino-acid pair predictability over time should be the most important observation we need to make, after we have developed this measure. Nevertheless, it is meaningful to observe the behavior of something that is alive, otherwise it will not change along the time course.

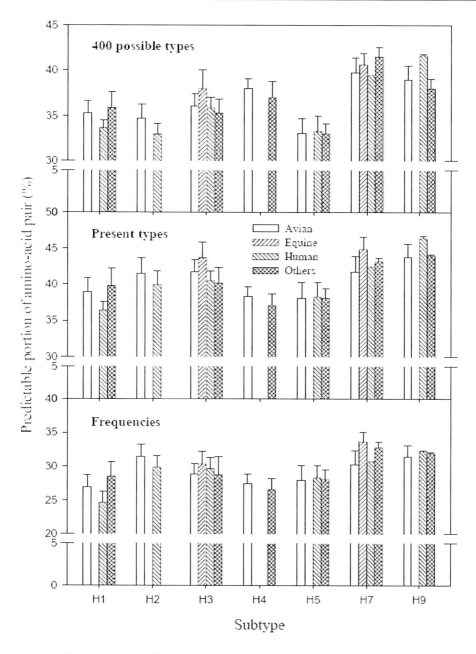

Figure 3-4. Predictable portions of hemagglutinins from influenza A virus cross species (the number of hemagglutinins in each species and subtype can be found in Table 3-6). Data are presented as mean±SD.

Being different from spatial angle, we need to set several different time points to observe the same object of interests. For a protein, we can consider it unchanged in its primary structure if no mutation occurs. Therefore, we can set our observation at two different time points, that is, before and after mutation.

3.2.1. Amino-Acid Pairs before and after Mutation

To understand, we analyze a protein being somewhat short but having many mutations recorded. This protein is the human Cx32 protein (gap junction beta-1 protein, accession number P08034). Actually, we have no particular aim in mind in choosing this protein or other proteins as examples in this book, and the only aim is easy for calculation, therefore the function of these proteins will not be our topics for discussion.

The human Cx32 protein is composed of 283 amino acids (Table 3-7) with 188 missense point mutations documented.

Table 3-7. Amino-acid composition of human Cx32 protein

Amino acid	Number before mutation	Number after mutation
		Position 3 "W" → "R"
Alanine, A	17	--
Arginine, R	19	20
Asparagine, N	9	--
Aspartic acid, D	8	--
Cysteine, C	10	--
Glutamic Acid, E	13	--
Glutamine, Q	8	--
Glycine, G	17	--
Histidine, H	10	
Isoleucine, I	17	--
Leucine, L	31	--
Lysine, K	12	--
Methionine, M	7	--
Phenylalanine, F	13	--
Proline, P	12	--
Serine, S	23	--
Threonine, T	12	--
Tryptophan, W	6	5
Tyrosine, Y	10	--
Valine, V	29	--

--, no change before and after mutation.

From the beginning of this protein, the first amino acid affected by mutation is tryptophan "W" located at position 3. As an amino-acid pair is related to two amino acids, the amino acids at positions 2 and 4 can construct two amino-acid pairs with "W", in this case, "NW" and "WT" because asparagine "N" is at position 2 and threonine "T" is at position 4.

At position 3, two mutations were recorded, "W" mutates to (i) arginine "R" or (ii) serine "S". Let us see what happens for the mutation changing "W" to "R". First, "NW" and "WT" will disappear, but second, "NR" and "RT" will appear. Thus, a point mutation will affect four amino-acid pairs in this case, i.e. "NW", "WT", "NR" and "RT", and we need to calculate these four amino-acid pairs before and after mutation to see their behaviors.

For the amino-acid pair predictability before and after mutation, we have the calculated results in Table 3-8. As can be seen, the actual and predicted frequencies in each amino-acid pair are changed before and after mutation, and the overall effect of mutation is to make the actual frequencies approach to their predicted ones.

Table 3-8. Amino-acid pairs before and after mutation in human Cx32 protein

Amino-acid pair	Before mutation			After mutation		
	Actual frequency	Predicted frequency	Attribute	Actual frequency	Predicted frequency	Attribute
NW	1	$9/283 \times 6/282 \times 282$ $= 0.1908 \approx 0$	IV	0	$9/283 \times 5/282 \times 282$ $= 0.1590 \approx 0$	II
WT	2	$6/283 \times 12/282 \times 282 = 0.2544 \approx 0$	IV	1	$5/283 \times 12/282 \times 282 = 0.2120 \approx 0$	IV
NR	1	$9/283 \times 19/282 \times 282 = 0.6042 \approx 1$	I	2	$9/283 \times 20/282 \times 282 = 0.6360 \approx 1$	III
RT	0	$19/283 \times 12/282 \times 282 = 0.8057 \approx 1$	V	1	$20/283 \times 12/282 \times 282 = 0.8481 \approx 1$	I

I, randomly predictable present type of amino-acid pair with predictable frequency; II, randomly predictable absent type of amino-acid pair; III, randomly predictable present type of amino-acid pair with unpredictable frequency; IV, randomly unpredictable present type of amino-acid pair; V, randomly unpredictable absent type of amino-acid pair.

3.2.2. Terminal Amino-Acid Pairs before and after Mutation

In the above example, we dealt with the mutation that does not occur at the terminals of a protein. In this example, we will look at the behavior of amino-acid pair predictability in the amino acid located at the terminal of a protein, at which a point mutation will directly affect only two amino-acid pairs.

There are several point mutations, which occur at the terminal position of human hemoglobin α-chain, where the amino acid is arginine "R". Four mutations are documented at this terminal position, changing "R" to cysteine "C" or leucine "L" or histidine "H" or proline "P". The amino acid next to the terminal position is tyrosine "Y", so the terminal amino-acid pair is "YR". A mutation, for example, making "R" to mutate to "C", will change two amino-acid pairs, "YR" and "YC".

Table 2-5 indicates that there were three "Y", three "R" and one "C" in human hemoglobin α-chain before mutation. After mutation, there are three "Y", two "R" and two "C". Thus, we have the following result in Table 3-9. In this table, the effect of this mutation seems to produce a balanced effect, i.e. changing one amino-acid pair from predictable to unpredictable, but vice versa in the other pair.

3.2.3. All Original and Mutated Amino-Acid Pairs before and after Mutation

In the above two examples, we have seen the changed behavior of amino-acid pair predictability related to a single point mutation, which affects either two or four amino-acid

Table 3-9. Amino-acid pairs before and after mutation in human hemoglobin α-chain

Amino-acid pair	Before mutation			After mutation		
	Actual frequency	Predicted frequency	Attribute	Actual frequency	Predicted frequency	Attribute
YR	1	$3/142 \times 3/141 \times 141$ $= 0.0634 \approx 0$	IV	0	$3/142 \times 2/141 \times 141 = 0.0423 \approx 0$	II
YC	0	$3/142 \times 1/141 \times 141$ $= 0.0211 \approx 0$	II	1	$3/142 \times 2/141 \times 141 = 0.0423 \approx 0$	IV

II, randomly predictable absent type of amino-acid pair; IV, randomly unpredictable present type of amino-acid pair.

pairs. Actually, the effect of a single point mutation is not limited in such a narrow range, i.e. two or four amino-acid pairs, because a single point mutation decreases the number of amino acids, one of which disappears after mutation, and increase the number of amino acids, one of which appears after mutation. Therefore a single point mutation can change the amino-acid composition of the protein, leading to the change in the predicted frequency of the amino-acid pairs that contain original and mutated amino acids.

It is very plausible that a single point mutation has its effect on the whole protein, because a protein is a functioning unit and any slight change would influence the whole protein. This is an important advantage of our method over others, so that we can estimate the overall effect of a single point mutation on the whole protein rather than two amino acids, original and mutated.

Let us see how a single point mutation change the predicted frequency in all amino-acid pairs containing original and mutated amino acids. In Table 3-10, we have seen that the mutation at position 3 of human Cx32 protein changes tryptophan "W" to arginine "R", as the consequence, the number of "R" increases to 20 from 19 while the number of "W" decreases to 5 from 6.

Table 3-10 lists the predicted frequencies in amino-acid pairs containing "W" before and after mutation. We can see that each predicted frequency is different before and after mutation. In principle, a single point mutation will lead the predicted frequency to change in 78 (39 + 39) amino-acid pairs. Meanwhile, Table 3-11 shows the predicted frequency in amino-acid pairs containing "R" before and after mutation, where we can also see that all the predicted frequencies vary before and after mutation. We can find the statistically significant difference in the predicted frequencies before and after mutation, using a statistical test, such as the student's t-test. Although the difference is small before and after mutation, it is highly different because all of the predicted frequencies move in the same direction, i.e. all predicted frequencies decrease in the amino-acid pairs, one of which disappears after mutation (Table 3-10) whereas all predicted frequencies increase in the amino-acid pairs, one of which appears after mutation (Table 3-11).

Table 3-10. Actual and predicted frequencies in amino-acid pairs containing "W" in human Cx32 protein before and after mutation at position 3 changing "W" to "R"

Amino-acid pair	Before mutation		After mutation	
	Actual frequency	Predicted frequency	Actual frequency	Predicted frequency
AW	0	0.3604	0	0.3004
RW	0	0.4028	0	0.3534
NW	1	0.1908	0	0.1590
DW	0	0.1696	0	0.1413
CW	0	0.2120	0	0.1767
EW	0	0.2756	0	0.2297
QW	0	0.1696	0	0.1413
GW	0	0.3604	0	0.3004
HW	0	0.2120	0	0.1767
IW	0	0.3604	0	0.3004
LW	2	0.6572	2	0.5477
KW	0	0.2544	0	0.2120
MW	0	0.1484	0	0.1237
FW	0	0.2756	0	0.2297
PW	0	0.2544	0	0.2120
SW	0	0.4876	0	0.4064
TW	0	0.2544	0	0.2120
WW	1	0.1060	1	0.0707
YW	0	0.2120	0	0.1767
VW	2	0.6148	2	0.5124
WA	0	0.3604	0	0.3004
WR	0	0.4028	0	0.3534
WN	0	0.1908	0	0.1590
WD	0	0.1696	0	0.1413
WC	0	0.2120	0	0.1767
WE	0	0.2756	0	0.2297
WQ	0	0.1696	0	0.1413
WG	1	0.3604	1	0.3004
WH	0	0.2120	0	0.1767
WI	0	0.3604	0	0.3004
WL	1	0.6572	1	0.5477
WK	0	0.2544	0	0.2120
WM	0	0.1484	0	0.1237
WF	0	0.2756	0	0.2297
WP	0	0.2544	0	0.2120
WS	1	0.4876	1	0.4064
WT	2	0.2544	1	0.2120
WY	0	0.2120	0	0.1767
WV	0	0.6148	0	0.5124

Comparison of predicted frequencies before and after mutation is statistically significant difference (the paired student's t-test, $P < 0.001$).

Table 3-11. Actual and predicted frequencies in amino-acid pairs containing "R" in human Cx32 protein before and after mutation at position 3 changing "W" to "R"

Amino-acid pair	Before mutation		After mutation	
	Actual frequency (AF)	Predicted frequency (PF)	Actual frequency (AF)	Predicted frequency (PF)
RA	2	1.1413	2	1.2014
RR	3	1.2085	3	1.3428
RN	0	0.6042	0	0.6360
RD	0	0.5371	0	0.5654
RC	1	0.6714	1	0.7067
RE	0	0.8728	0	0.9187
RQ	0	0.5371	0	0.5654
RG	0	1.1413	0	1.2014
RH	2	0.6714	2	0.7067
RI	1	1.1413	1	1.2014
RL	5	2.0813	5	2.1908
RK	1	0.8057	1	0.8481
RM	0	0.4700	0	0.4947
RF	0	0.8728	0	0.9187
RP	1	0.8057	1	0.8481
RS	2	1.5442	2	1.6254
RT	0	0.8057	1	0.8481
RW	0	0.4028	0	0.3534
RY	0	0.6714	0	0.7067
RV	1	1.9470	1	2.0495
AR	1	1.1413	1	1.2014
NR	1	0.6042	2	0.6360
DR	1	0.5371	1	0.5654
CR	0	0.6714	0	0.7067
ER	0	0.8728	0	0.9187
QR	1	0.5371	1	0.5654
GR	1	1.1413	1	1.2014
HR	1	0.6714	1	0.7067
IR	1	1.1413	1	1.2014
LR	2	2.0813	2	2.1908
KR	1	0.8057	1	0.8481
MR	0	0.4700	0	0.4947
FR	2	0.8728	2	0.9187
PR	0	0.8057	0	0.8481
SR	2	1.5442	2	1.6254
TR	0	0.8057	0	0.8481
WR	0	0.4028	0	0.3534
YR	0	0.6714	0	0.7067
VR	2	1.9470	2	2.0495

Comparison of predicted frequencies before and after mutation is statistically significant difference (the paired student's t-test, $P < 0.001$).

3.2.4. Predictable Portion in Type and in Frequency

The calculation in the three examples above is very complicated and labor-extensive, although it shows the changing behavior of amino-acid pair predictability. Certainly, the energy-saving way is to calculate the predictable or unpredictable portion of amino-acid pairs before and after mutation.

In Section 2.6.6, we classify the predictable and unpredictable portions of amino-acid pairs in a protein into two categories, (i) the predictable and unpredictable portions in type, which refers to how many predictable types a protein has with respect to the 400 possible types of amino-acid pairs or with respect to the present types in the protein, and (ii) the predictable and unpredictable portions in frequency, which refers to how many pairs of all present types the protein has.

Now let us look at these concepts and their calculations in great details. Once again, we use human Cx32 protein in this analysis. We already know that human Cx32 protein contains 283 amino acids constructing 282 amino-acid pairs.

We count these 282 amino-acid pairs against 400 possible types of amino-acid pairs (Table 2-1) with the aims of determining how many types human Cx32 protein has with reference to 400 possible types, and how many amino-acid pairs each type has. Without any elaboration, we know that human Cx32 protein cannot contain all of 400 possible types of amino-acid pairs because it has only 282 amino-acid pairs, which are less than 400 types. The detailed count we conducted is as follows. Of 400 possible types of amino-acid pairs, 211 types do not appear in human Cx32 protein, indicating that this protein contains only 189 types of amino-acid pairs (400 − 211 = 189).

On the other hand, we know that human Cx32 protein has 282 amino-acid pairs, which are larger than 189 types of amino-acid pairs, so there must be some repetitions for some present types (Table 3-12).

Table 3-12. Counting amino-acid pairs of human Cx32 protein

Appearances	Number of types	Number of frequencies
0	211	$211 \times 0 = 0$
1	125	$125 \times 1 = 125$
2	47	$47 \times 2 = 94$
3	8	$8 \times 3 = 24$
4	6	$6 \times 4 = 24$
5	3	$3 \times 5 = 15$
Total	400	282

Looking at the details of 189 present types of amino-acid pairs, Table 3-13 lists all of the counts with respect to amino-acid pair predictability.

(i) We have $\Sigma(AF = 0) = 211$, which we have mentioned above.

(ii) We have $\Sigma(AF = PF) = 209$, which include both the randomly predictable present type of amino-acid pair with predictable frequency ($AF = PF > 0$, i.e. Attribute I) and randomly predictable absent type of amino-acid pair ($AF = PF = 0$, i.e. Attribute II).

(iii) We exclude $\Sigma(AF = PF = 0) = 125$ from $\Sigma(AF = PF) = 209$, and then we have the randomly predictable present type of amino-acid pair with predictable frequency (Attribute I), which are $\Sigma(AF = PF > 0) = 84$.

(iv) We know that the human Cx32 protein has 189 types of amino-acid pairs, thus the predictable portion in type (%) = 84/189 = 44.44%.

(v) If we count the frequency in each of 84 types of amino-acid pairs ($AF = PF > 0$), we find there are 95 amino-acid pairs in total. Meanwhile we know that the human Cx32 protein has 282 amino-acid pairs, so the predictable portion in frequency (%) = 95/282 = 33.69%.

(vi) On the other hand, we can find out which attribute a type of amino-acid pair belongs to. For instance, the type "AA" has $AF = 2$ but $PF = 1$, being the case that the randomly predictable present type of amino-acid pair with unpredictable frequency (Attribute III).

3.2.5. Predictable Portion with Respect to 400 Possible Types

In the above example, we see the detailed calculation of predictable portions in both type and frequency with respect to the 189 types of amino-acid pairs and 282 pairs in human Cx32 protein. This way, we can easily calculate the unpredictable portion because the sum of predictable and unpredictable portions is equal to 100%.

Additionally, one might note that the amino-acid pair predictability can successfully predict the number of absent types of amino-acid pairs, i.e. $AF = PF = 0$ (Attribute II), which however is excluded in the above calculation of predictable portion where we consider only the types and pairs present in human Cx32 protein.

Therefore we can also calculate the predictable portion including the predictable absent types, which although are not located in the protein. In this case, our reference should be directed to the 400 possible types rather than the types present in the protein, so the predictable portion in type is equal to 52.25% for human Cx32 protein, i.e. $\Sigma(AF = PF > 0)$ + $\Sigma(AF = PF = 0) = 84 + 125 = 209$, then 209/400 = 52.25(%), which is different from the predictable portion 44.44% in the previous worked example. Actually, we have used the predictable portion with respect to the 400 possible types in a number of studies because the absent types may appear after mutations although they do not exist in the protein before mutations [26, 29-47, 63, 64, 68-73].

In fact, Figures 3-1, 3-2, 3-3 and 3-4 already show the difference between three calculations of predictable portions.

Table 3-13. Amino-acid pairs in human Cx32 protein with respect to 400 possible types of amino-acid pairs

Type	AF	PF	Type	AF	PF	Type	AF	PF	Type	AF	PF	Type	AF	PF	Type	AF	PF
AA	2	1	RA	2	1	NA	0	1	DA	0	0	CA	1	0	EA	1	1
AR	1	1	RR	3	1	NR	1	1	DR	1	1	CR	0	1	ER	0	1
AN	0	1	RN	0	1	NN	0	0	DN	0	0	CN	2	0	EN	0	0
AD	0	0	RD	0	0	ND	0	0	DD	0	0	CD	1	0	ED	0	0
AC	2	1	RC	1	1	NC	0	0	DC	1	0	CC	0	0	EC	0	0
AE	3	1	RE	0	1	NE	1	0	DE	1	0	CE	0	0	EE	1	1
AQ	1	0	RQ	0	1	NQ	0	0	DQ	1	0	CQ	0	0	EQ	1	0
AG	1	1	RG	0	1	NG	0	1	DG	1	0	CG	0	1	EG	1	1
AH	1	1	RH	2	1	NH	0	0	DH	0	0	CH	0	0	EH	0	0
AI	1	1	RI	1	1	NI	0	1	DI	1	0	CI	1	1	EI	1	1
AL	1	2	RL	5	2	NL	0	1	DL	0	1	CL	0	1	EL	0	1
AK	0	1	RK	1	1	NK	1	0	DK	0	0	CK	0	0	EK	4	1
AM	2	0	RM	0	0	NM	0	0	DM	0	0	CM	0	0	EM	0	0
AF	0	1	RF	0	1	NF	0	0	DF	0	0	CF	1	0	EF	0	1
AP	0	1	RP	1	1	NP	1	0	DP	1	0	CP	1	0	EP	0	1
AS	1	1	RS	2	2	NS	1	1	DS	0	1	CS	1	1	ES	1	1
AT	0	1	RT	0	1	NT	2	0	DT	0	0	CT	0	0	ET	0	1
AW	0	0	RW	0	0	NW	1	0	DW	0	0	CW	0	0	EW	0	0
AY	0	1	RY	0	1	NY	0	0	DY	0	0	CY	1	0	EY	1	0
AV	1	2	RV	1	2	NV	1	1	DV	1	1	CV	0	1	EV	2	1
AF=0	8			10			12			12			11			11	
AF=PF	7			8			11			12			10			11	
AF=PF=0	2			3			8			10			8			6	
AF=PF>0 in type	5			5			3			2			2			5	
AF=PF>0 in frequency	5			6			3			2			2			5	

Table 3-13. (Continued)

Type	AF	PF	Type	AF	PF	Type	AF	PF	Type	AF	PF	Type	AF	PF	Type	AF	PF	Type	AF	PF
QA	0	0	GA	1	1	HA	0	1	IA	0	1	LA	2	2	KA	0	1	MA	0	0
QR	1	1	GR	1	1	HR	1	1	IR	1	1	LR	2	2	KR	1	1	MR	0	0
QN	1	0	GN	0	1	HN	0	0	IN	1	0	LN	1	1	KN	0	0	MN	1	0
QD	1	0	GD	2	0	HD	0	0	ID	0	0	LD	0	1	KD	1	0	MD	0	0
QC	0	0	GC	1	1	HC	0	0	IC	2	1	LC	0	1	KC	1	0	MC	0	0
QE	0	0	GE	0	1	HE	0	0	IE	0	1	LE	2	1	KE	0	1	ME	0	0
QQ	1	0	GQ	0	0	HQ	1	1	IQ	0	0	LQ	2	1	KQ	1	0	MQ	0	0
QG	0	0	GG	0	1	HG	1	1	IG	1	1	LG	0	2	KG	1	1	MG	0	0
QH	1	0	GH	2	1	HH	0	0	IH	0	1	LH	1	1	KH	0	0	MH	1	0
QI	0	0	GI	1	1	HI	2	1	II	2	1	LI	2	2	KI	0	1	MI	0	0
QL	1	1	GL	2	2	HL	1	1	IL	3	1	LL	5	3	KL	1	1	ML	2	1
QK	0	0	GK	0	1	HK	1	0	IK	0	1	LK	1	1	KK	1	0	MK	0	0
QM	0	0	GM	0	0	HM	0	0	IM	1	1	LM	0	1	KM	1	1	MM	0	0
QF	1	0	GF	1	1	HF	0	0	IF	2	0	LF	1	1	KF	0	0	MF	0	0
QP	1	0	GP	0	1	HP	0	0	IP	0	1	LP	0	1	KP	0	1	MP	0	0
QS	0	1	GS	2	1	HS	1	0	IS	3	1	LS	4	3	KS	2	1	MS	0	1
QT	0	0	GT	2	1	HT	0	0	IT	0	0	LT	0	1	KT	1	1	MT	0	0
QW	0	0	GW	0	0	HW	0	0	IW	0	1	LW	2	1	KW	0	0	MW	0	0
QY	0	0	GY	1	1	HY	0	0	IY	1	1	LY	2	1	KY	0	1	MY	1	0
QV	0	1	GV	1	2	HV	2	1	IV	0	2	LV	4	3	KV	1	0	MV	2	1
AF=0	12			8			12			10			6			9			15	
AF=PF	12			10			15			7			7			9			14	
AF=PF=0	10			3			11			3			0			4			14	
AF=PF>0 in type	2			7			4			4			7			5			0	
AF=PF>0 in frequency	2			8			4			4			10			5			0	

Type	AF	PF	Type	AF	PF	Type	AF	PF	Type	AF	PF	Type	AF	PF	Type	AF	PF	Type	AF	PF
FA	0	1	PA	1	1	SA	1	1	TA	1	1	WA	0	1	YA	1	1	VA	4	2
FR	2	1	PR	0	1	SR	2	2	TR	0	2	WR	0	1	YR	0	1	VR	2	2
FN	0	0	PN	1	0	SN	1	1	TN	0	1	WN	0	0	YN	0	0	VN	1	1
FD	0	0	PD	0	0	SD	1	1	TD	0	1	WD	0	0	YD	1	0	VD	1	1
FC	0	0	PC	1	0	SC	0	1	TC	0	1	WC	0	0	YC	0	0	VC	1	1
FE	1	1	PE	1	1	SE	1	1	TE	1	1	WE	0	1	YE	0	0	VE	0	1
FQ	0	0	PQ	0	0	SQ	0	1	TQ	0	1	WQ	0	0	YQ	0	0	VQ	0	1
FG	1	1	PG	3	1	SG	4	1	TG	2	1	WG	1	1	YG	1	1	VG	0	2
FH	0	0	PH	0	0	SH	1	1	TH	0	1	WH	0	0	YH	0	0	VH	1	1
FI	2	1	PI	1	1	SI	0	1	TI	0	1	WI	0	1	YI	0	1	VI	2	1
FL	0	1	PL	1	1	SL	2	3	TL	3	3	WL	1	1	YL	2	1	VL	1	2
FK	0	1	PK	0	1	SK	0	1	TK	0	1	WK	0	1	YK	1	0	VK	2	3
FM	2	0	PM	0	0	SM	0	1	TM	0	1	WM	0	0	YM	0	0	VM	0	1
FF	1	1	PF	0	1	SF	1	1	TF	0	1	WF	0	1	YF	0	0	VF	5	1
FP	1	1	PP	1	0	SP	2	1	TP	1	1	WP	0	1	YP	2	0	VP	0	1
FS	0	1	PS	1	1	SS	1	2	TS	0	1	WS	1	1	YS	0	1	VS	2	2
FT	1	1	PT	1	1	ST	2	1	TT	0	0	WT	2	0	YT	1	0	VT	0	1
FW	0	0	PW	0	0	SW	0	0	TW	0	0	WW	1	0	YW	0	0	VW	2	1
FY	1	0	PY	0	0	SY	0	1	TY	1	1	WY	0	0	YY	0	0	VY	2	1
FV	1	1	PV	0	1	SV	4	2	TV	3	3	WV	0	1	YV	2	1	VV	3	3
AF=0	10			10			7			13			15			13			6	
AF=PF	12			12			8			11			15			10			8	
AF=PF=0	6			6			1			8			14			9			0	
AF=PF>0 in type	6			6			7			3			1			1			8	
AF=PF>0 in frequency	6			6			8			3			1			1			13	

3.2.6. Predictable Portion before and after Mutation

In the above two examples, we have conducted the detailed counting as well as the calculation of predictable portion in type and in frequency with respect to either the present types or 400 possible types. Now we investigate how the predictable portion behaves before and after mutation in human Cx32 protein, as we have shown how a single point mutation affects the amino-acid pairs containing original and mutated amino acids in Section 3.2.3.

Table 3-14 shows all 188 missense point mutations in human Cx32 protein and their effects on the predictable portions of amino-acid pairs. In this table, we list the predictable portions discussed in Sections 3.2.4 and 3.2.5: (i) the predictable portion in 400 possible types of amino-acid pairs in human Cx32 protein (52.25%), (ii) the predictable portion in present types (44.44%), and (iii) the predictable portion in frequencies (33.69%).
We can see the predictable portion increased, or unchanged or decreased after mutations in the last column of Table 3-14. For the predicable portion in 400 possible types of amino-acid pairs (column 3 in Table 3-14), there are 54 (28.72%) increased, 26 (13.83%) unchanged and 108 (57.45%) decreased. For the predicable portion in present types in the protein (column 5 in Table 3-14), there are 78 (41.49%) increased, 24 (12.77%) unchanged and 86 (45.74%) decreased. With respect to the predictable portion in frequency (column 7 in Table 3-14), there are 64 (34.04%) increased, 32 (17.02%) unchanged and 92 (48.94%) decreased.

This way, we have observed the behavior of amino-acid pair predictability in human Cx32 protein from the time angle, and we can conclude that the amino-acid pair predictability passed the test.

Table 3-14. 188 missense point mutations in human Cx32 protein and their effect on predictable portions of amino-acid pairs

Position	Mutation	Predictable portion in 400 possible types (%)	Effect	Predictable portion in present types (%)	Effect	Predictable portion in frequency (%)	Effect
Normal Cx32 protein		52.25		44.44		33.69	
3	W → R	52.50	↑	44.44	↔	33.69	↔
3	W → S	52.25	↔	44.15	↓	33.33	↓
7	Y → C	52.25	↔	45.79	↑	34.75	↑
8	T → I	51.75	↓	43.92	↓	33.33	↓
8	T → P	52.50	↑	44.15	↓	33.33	↓
9	L → W	52.00	↓	44.74	↑	34.40	↑
11	S → G	52.00	↓	44.50	↑	33.69	↔
12	G → S	52.50	↑	45.74	↑	34.75	↑
13	V → L	51.50	↓	43.32	↓	32.27	↓
13	V → M	51.25	↓	44.15	↓	33.33	↓
14	N → K	52.75	↑	43.32	↓	32.62	↓
15	R → Q	50.50	↓	43.92	↓	32.62	↓
15	R → W	51.50	↓	43.92	↓	32.62	↓
16	H → P	52.00	↓	44.15	↓	33.33	↓

Position	Mutation	Predictable portion in 400 possible types (%)	Effect	Predictable portion in present types (%)	Effect	Predictable portion in frequency (%)	Effect
20	I → S	51.50	↓	43.85	↓	32.98	↓
21	G → D	51.50	↓	44.68	↑	33.69	↔
22	R → G	51.75	↓	44.68	↑	33.33	↓
22	R → P	52.25	↔	44.44	↔	32.98	↓
22	R → Q	51.00	↓	44.44	↔	32.98	↓
23	V → A	51.75	↓	45.50	↑	34.38	↑
24	W → C	51.75	↓	44.97	↑	34.04	↑
25	L → F	52.25	↔	44.74	↑	34.40	↑
25	L → P	52.00	↓	44.44	↔	34.04	↑
26	S → L	52.00	↓	43.92	↓	32.98	↓
26	S → W	52.25	↔	44.74	↑	33.69	↔
28	I → N	51.75	↓	43.68	↓	32.98	↓
28	I → T	52.50	↑	45.03	↑	34.05	↑
29	F → L	52.25	↔	44.44	↔	33.33	↓
30	I → N	52.25	↔	45.03	↑	34.40	↑
30	I → T	52.50	↑	45.26	↑	34.40	↑
34	M → I	52.75	↑	45.50	↑	34.40	↑
34	M → K	53.50	↑	45.50	↑	34.40	↑
34	M → T	53.25	↑	45.50	↑	34.40	↑
34	M → V	53.00	↑	44.44	↔	32.98	↓
35	V → M	51.75	↓	44.97	↑	34.04	↑
37	V → M	52.00	↓	44.74	↑	34.04	↑
38	V → M	51.50	↓	43.98	↓	32.98	↓
39	A → P	52.75	↑	45.26	↑	34.40	↑
39	A → V	52.25	↔	44.44	↔	32.98	↓
40	A → T	52.25	↔	44.74	↑	34.04	↑
40	A → V	53.00	↑	45.79	↑	35.11	↑
41	E → K	52.25	↔	44.44	↔	33.69	↔
43	V → M	52.00	↓	45.03	↑	34.40	↑
44	W → L	52.75	↑	44.97	↑	34.04	↑
49	S → P	52.50	↑	44.97	↑	33.69	↔
49	S → Y	52.25	↔	45.79	↑	34.40	↑
50	S → P	52.25	↔	44.68	↑	33.33	↓
53	C → S	52.25	↔	44.97	↑	34.04	↑
55	T → A	51.75	↓	44.21	↓	34.04	↑
55	T → I	51.50	↓	43.68	↓	33.33	↓
55	T → R	51.50	↓	42.33	↓	32.27	↓
56	L → F	52.75	↑	45.55	↑	35.11	↑
57	Q → H	51.25	↓	43.92	↓	33.33	↓
58	P → R	52.50	↑	43.39	↓	32.98	↓

Table 3-14. Continued

Position	Mutation	Predictable portion in 400 possible types (%)	Effect	Predictable portion in present types (%)	Effect	Predictable portion in frequency (%)	Effect
59	G → C	51.25	↓	44.44	↔	33.69	↔
59	G → R	52.00	↓	43.92	↓	33.33	↓
60	C → F	51.25	↓	42.86	↓	32.62	↓
63	V → I	51.75	↓	44.97	↑	34.04	↑
64	C → F	52.00	↓	43.85	↓	32.98	↓
64	C → S	52.25	↔	44.68	↑	33.33	↓
65	Y → C	52.00	↓	45.21	↑	34.04	↑
65	Y → H	51.25	↓	44.44	↔	33.69	↔
69	F → L	52.50	↑	44.44	↔	33.69	↔
70	P → A	51.25	↓	43.09	↓	32.62	↓
75	R → P	51.75	↓	43.16	↓	31.91	↓
75	R → Q	50.50	↓	43.16	↓	31.91	↓
75	R → W	51.25	↓	42.86	↓	31.56	↓
77	W → S	53.00	↑	45.74	↑	34.75	↑
80	Q → R	52.00	↓	43.62	↓	32.62	↓
81	L → F	52.00	↓	44.15	↓	33.33	↓
83	L → P	53.75	↑	47.12	↑	37.23	↑
84	V → I	51.50	↓	44.44	↔	33.69	↔
85	S → C	51.00	↓	43.68	↓	32.62	↓
85	S → F	51.50	↓	42.86	↓	31.91	↓
86	T → A	51.75	↓	43.92	↓	33.33	↓
86	T → N	52.00	↓	43.09	↓	32.62	↓
86	T → S	52.50	↑	44.68	↑	34.04	↑
87	P → A	51.50	↓	43.32	↓	32.62	↓
87	P → L	52.00	↓	42.78	↓	31.91	↓
87	P → S	52.25	↔	43.62	↓	32.98	↓
89	L → P	52.75	↑	45.50	↑	34.75	↑
90	L → H	51.50	↓	44.97	↑	35.11	↑
90	L → V	52.25	↔	44.44	↔	33.33	↓
91	V → M	52.00	↓	45.03	↑	35.11	↑
93	M → V	53.00	↑	44.68	↑	34.04	↑
94	H → D	52.00	↓	45.50	↑	34.40	↑
94	H → Q	51.75	↓	45.79	↑	34.75	↑
94	H → Y	52.75	↑	46.28	↑	34.75	↑
95	V → M	51.50	↓	44.50	↑	34.04	↑
100	H → Y	52.75	↑	46.32	↑	35.11	↑
102	E → G	51.50	↓	43.92	↓	33.33	↓
103	K → E	53.00	↑	44.15	↓	33.69	↔
104	K → T	52.50	↑	44.15	↓	33.33	↓

Position	Mutation	Predictable portion in 400 possible types (%)	Effect	Predictable portion in present types (%)	Effect	Predictable portion in frequency (%)	Effect
107	R → W	51.25	↓	42.86	↓	31.56	↓
108	L → P	52.50	↑	44.97	↑	34.40	↑
120	V → E	52.75	↑	45.50	↑	34.75	↑
124	K → N	52.25	↔	43.09	↓	32.62	↓
125	V → D	51.25	↓	44.68	↑	33.69	↔
126	H → Y	52.50	↑	46.03	↑	34.75	↑
127	I → M	52.00	↓	45.03	↑	34.40	↑
127	I → S	52.50	↑	45.50	↑	34.75	↑
128	S → P	52.50	↑	44.74	↑	33.69	↔
130	T → I	51.25	↓	43.39	↓	32.98	↓
131	L → P	52.50	↑	45.26	↑	34.75	↑
133	W → C	51.50	↓	44.74	↑	34.04	↑
133	W → R	52.50	↑	44.74	↑	34.04	↑
135	Y → C	52.25	↔	45.79	↑	34.75	↑
136	V → A	51.25	↓	43.92	↓	32.98	↓
138	S → N	51.50	↓	42.86	↓	32.27	↓
139	V → M	51.75	↓	44.21	↓	32.98	↓
141	F → L	52.00	↓	43.92	↓	32.98	↓
142	R → E	52.00	↓	43.68	↓	32.98	↓
142	R → Q	50.50	↓	43.68	↓	32.62	↓
142	R → W	51.50	↓	43.68	↓	32.62	↓
146	E → K	52.25	↔	44.44	↔	33.69	↔
147	A → D	51.25	↓	44.68	↑	33.69	↔
149	F → I	51.25	↓	43.92	↓	32.98	↓
149	F → V	52.00	↓	43.68	↓	32.62	↓
151	Y → S	53.00	↑	46.03	↑	34.75	↑
153	F → S	52.25	↔	44.44	↔	33.33	↓
156	L → F	52.50	↑	44.97	↑	34.40	↑
156	L → R	52.75	↑	45.26	↑	34.40	↑
157	Y → C	52.25	↔	45.79	↑	34.75	↑
158	P → A	51.25	↓	42.86	↓	32.62	↓
158	P → R	52.75	↑	43.98	↓	33.69	↔
158	P → S	52.50	↑	44.21	↓	33.69	↔
159	G → D	51.25	↓	44.74	↑	34.04	↑
159	G → S	52.00	↓	44.44	↔	33.69	↔
160	Y → H	51.00	↓	44.15	↓	33.33	↓
161	A → P	52.00	↓	44.44	↔	33.69	↔
164	R → Q	50.50	↓	43.16	↓	31.91	↓
164	R → W	51.25	↓	42.86	↓	31.56	↓
168	C → R	52.75	↑	44.68	↑	33.69	↔

Table 3-14. Continued

Position	Mutation	Predictable portion in 400 possible types (%)	Effect	Predictable portion in present types (%)	Effect	Predictable portion in frequency (%)	Effect
168	C → Y	52.50	↑	45.74	↑	34.40	↑
172	P → A	51.75	↓	43.62	↓	32.98	↓
172	P → L	52.75	↑	43.92	↓	33.33	↓
172	P → R	52.50	↑	43.39	↓	32.98	↓
172	P → S	53.00	↑	44.74	↑	34.04	↑
173	C → R	52.75	↑	44.68	↑	33.69	↔
175	N → D	52.75	↑	44.74	↑	34.04	↑
177	V → A	51.50	↓	44.44	↔	33.69	↔
177	V → E	51.75	↓	43.92	↓	33.69	↔
178	D → Y	52.00	↓	44.68	↑	33.69	↔
179	C → R	52.75	↑	44.68	↑	33.69	↔
180	F → L	52.25	↔	44.15	↓	33.33	↓
180	F → S	51.75	↓	43.85	↓	32.98	↓
181	V → A	51.25	↓	43.92	↓	32.98	↓
181	V → M	51.50	↓	43.92	↓	32.98	↓
182	S → T	52.25	↔	44.50	↑	33.33	↓
183	R → C	51.50	↓	44.44	↔	32.98	↓
183	R → H	50.25	↓	42.86	↓	31.91	↓
183	R → S	52.00	↓	44.68	↑	33.69	↔
184	P → L	52.50	↑	43.62	↓	32.98	↓
184	P → R	52.50	↑	43.62	↓	32.98	↓
186	E → K	52.00	↓	43.92	↓	33.33	↓
187	K → E	53.25	↑	44.44	↔	34.04	↑
189	V → G	51.75	↓	44.44	↔	33.69	↔
189	V → I	52.00	↓	45.26	↑	34.40	↑
191	T → A	51.50	↓	43.92	↓	33.69	↔
192	V → F	52.00	↓	43.68	↓	33.33	↓
193	F → C	51.00	↓	43.16	↓	32.98	↓
193	F → L	52.25	↔	44.21	↓	33.69	↔
194	M → V	52.75	↑	44.44	↔	33.69	↔
198	S → F	51.50	↓	42.86	↓	32.27	↓
199	G → R	51.50	↓	43.09	↓	32.27	↓
201	C → R	51.75	↓	43.62	↓	32.98	↓
201	C → Y	52.75	↑	46.32	↑	35.11	↑
203	I → N	52.75	↑	45.26	↑	34.75	↑
204	L → F	52.25	↔	44.97	↑	34.75	↑
204	L → V	52.25	↔	44.44	↔	34.40	↑
205	N → I	52.00	↓	43.62	↓	32.62	↓
205	N → S	52.75	↑	43.85	↓	32.98	↓

Position	Mutation	Predictable portion in 400 possible types (%)	Effect	Predictable portion in present types (%)	Effect	Predictable portion in frequency (%)	Effect
208	E → G	52.00	↓	44.97	↑	34.40	↑
208	E → K	52.50	↑	44.74	↑	34.04	↑
211	Y → H	51.25	↓	44.44	↔	33.69	↔
213	I → V	52.00	↓	43.92	↓	32.62	↓
215	R → Q	51.25	↓	44.74	↑	33.33	↓
215	R → W	51.25	↓	43.68	↓	32.62	↓
219	R → C	51.25	↓	43.92	↓	32.62	↓
219	R → H	50.00	↓	42.02	↓	31.21	↓
220	R → G	51.75	↓	44.21	↓	32.98	↓
230	R → C	51.25	↓	44.21	↓	32.98	↓
230	R → L	51.75	↓	43.62	↓	32.98	↓
235	F → C	51.00	↓	43.09	↓	32.62	↓
238	R → H	50.00	↓	42.33	↓	31.56	↓
239	L → I	51.75	↓	44.97	↑	34.40	↑
264	R → C	51.50	↓	43.98	↓	32.62	↓
280	C → G	51.75	↓	44.68	↑	33.69	↔

3.3. Behavior of Amino-Acid Pair Predictability from Spatial and Time Angles

When observing the behavior of amino-acid pair predictability from spatial angle, we mainly deal with multiple proteins. By contrast, we mainly deal with a single protein when observing the behavior of amino-acid pair predictability from time angle. Each observation has its advantage and limitation.

The behavior of amino-acid pair predictability from both spatial and time angles, on the other hand, suggests that the scale of obversation would be large because a single protein cannot make a spatial observation. Thus, this observation is more suitable for analyzing the trend of a protein family over time, or what we observe is the evolutionary process of proteins of interests. In particular, we need to observe the behavior of predictable portion, which is suitable for a whole protein, in proteins at different time points. Still we use the dataset containing 2495 hemagglutinins from influenza A viruses (Tables 3-5 and 3-6) in our observation. Table 3-15 lists these 2495 hemagglutinins grouped according to their isolated year.

Figure 3-5 displays the predictable portions of influenza A virus hemagglutinins from 1918 to 2006. In this figure, a black symbol and capped gray bar are the mean and SD for all hemagglutinins in the given year. First, we can see the predictable portion different one another, which answers our question of whether the amino-acid pair predictability changes over time and space. Second, the regressed lines indicate the trend of predictable portion along the time course for a protein family, that is, the predictable portion increases over time. This fully supports our notation that the parsimony of nature makes the predictable portion increase through randomness (for review see [39, 63, 64]).

Table 3-15. 2495 hemagglutinins grouped according to year of isolation

Year	No.	Year	No.	Year	No.	Year	No	Year	No.	Year	No.
1918	1	1946	1	1963	9	1974	9	1985	33	1996	73
1927	2	1947	2	1964	3	1975	10	1986	20	1997	87
1930	2	1949	2	1965	3	1976	28	1987	20	1998	93
1931	1	1950	1	1966	7	1977	45	1988	23	1999	178
1933	6	1954	2	1967	2	1978	18	1989	11	2000	201
1934	6	1956	7	1968	15	1979	28	1990	8	2001	185
1935	1	1957	13	1969	2	1980	27	1991	20	2002	206
1939	1	1958	6	1970	5	1981	11	1992	14	2003	265
1940	1	1959	2	1971	13	1982	14	1993	16	2004	279
1942	1	1960	1	1972	20	1983	49	1994	29	2005	218
1943	1	1961	3	1973	9	1984	32	1995	49	2006	45

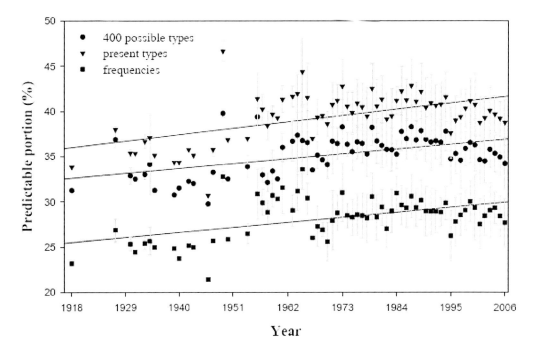

Figure 3-5. Predictable portions of influenza A virus hemagglutinins from 1918 to 2006. The data are presented as mean±SD. The solid lines are regressed lines for given predictable portion.

From research viewpoint, Figure 3-5 is very suitable for us to analyze the mutation trend of a protein family along the time course.

3.4. Meaning of Amino-Acid Pair Predictability

After observing the behavior of amino-acid pair predictability from spatial and time angles, we are more familiar with the amino-acid pair predictability in different scenarios. We can therefore figure out the meanings of amino-acid pair predictability. Here, our approach is somewhat different from other approaches, whose meanings were defined before observing the behaviors.

3.4.1. General Meanings

In calculation, we see that the permutation gives the frequency, which is the biggest chance of occurrence of amino-acid pair in question. This predicted frequency means that the amino-acid pair can be constructed with the least time and energy. Logically, nature should do so if it has no particular reason to spend more time and energy to construct the unpredictable present amino-acid pairs, or deliberately does not allow to construct certain amino-acid pairs, the unpredictable absent amino-acid pairs, in a protein.

On the other hand, the spending of the least time and energy and the use of available resource are the optimal condition for catching up with the fast evolutionary process, and a species would be extinct without the parsimony.

Along this line of thought, we can score the amino-acid pair predictability according to its attributes (Table 3-16).

3.4.2. Randomness and Mutation

The examples shown in this section and others we studied in the past demonstrate that the amino-acid pair predictability can reliably record the change in a protein induced by mutations [29-47, 68-73]. In Figure 3-5, we have seen the trend of influenza A virus hemagglutinins that the predictable portion increases with fluctuation along the time course. Likely, the protein or nature feels uncomfortable and uneasy with so many unpredictable present amino-acid pairs, thus it reduces the unpredictable portion through mutation. However, the result again creates new unpredictable present amino-acid pairs, which nature still feels uncomfortable and needs to be mutated out. This may partially explain the reasoning of continuing mutations.

Meanwhile, we have seen that more than one third of 188 point mutations in human Cx32 protein can increase the predictable portion of amino-acid pairs. Hence, we can say that the amino-acid pair predictability not only records the any change in proteins, but also indicates the percentage of mutations engineered randomly.

Table 3-16. Score of amino-acid pair predictability

Attributes		Meaning	Score
I	Randomly predictable present type of amino-acid pair with predictable frequency	The type may be useful with the least time and energy for construction.	+
III	Randomly predictable present type of amino-acid pair with unpredictable frequency	Both the type and the different frequency are necessary for the protein.	++
IV	Randomly unpredictable present type of amino-acid pair	The type is absolutely necessary for the protein.	+++
II	Randomly predictable absent type of amino-acid pair	The type is useless for the protein.	-
V	Randomly unpredictable absent type of amino-acid pair	The type is harmful for the protein.	--
I II	Predictable portion	The larger the predictable portion is, the less time and energy the need for protein construction is, so the protein is more likely to be constructed randomly.	
III IV V	Unpredictable portion	The larger the unpredictable portion is, the more time and energy the need for protein construction is, thus the protein is more likely to be constructed deliberately.	

3.4.3. Implication for Research

In general, the amino-acid pair predictability indicates how easy or difficult to construct an amino-acid pair as well as a whole protein sequence, so what this meaning implies? It implies the power engineering the protein evolution through mutation. Thus, we can use the amino-acid predictability as a measure to determine the mutation trend in a protein because the demand of randomness will constantly work hard to minimize the unpredictable portion through mutation. However, each mutation changes the amino-acid composition in the protein, which results in new unpredicted types of amino-acid pairs engineering new mutation in future. As the presence of unpredictable portion continues, the mutation will continue and the evolution will continue.

Hence, the amino-acid predictability implicates the history, current state and future of a protein (for review see [39, 63, 64]). By comparison of unpredictable with predictable portions, we can estimate the history of a protein, because we expect to see the decrease in the unpredictable portion in a protein as protein evolution goes on through mutation. Similarly, this comparison also provides us with the estimation of the current state and the future of proteins.

Chapter 4

Application of Amino-Acid Pair Predictability

After determined that the amino-acid pair predictability is a dynamic measure rather than a constant one in spatial and time axes, we can apply it to analyzing proteins and also we can apply various mathematical models to analyzing proteins.

In this Chapter, we will mainly discuss the application of amino-acid pair predictability to the research, thus this Chapter is more research-oriented. Perhaps this Chapter is more suitable for the researchers rather than university students.

At this stage, we apply the amino-acid pair predictability at two levels, that is, micro-level and macro-level. At micro-level, we analyze the concrete amino-acid pairs, while we analyze the general trend at macro-level.

4.1. Amino-Acid Pairs Sensitive to Mutation

An important application of amino-acid pair predictability is to find the amino-acid pairs sensitive to mutation in a protein. Nevertheless, currently there are many methods, which are trying to find the amino acid sensitive to mutations in a protein [74-80]. Of course, each method is based on its own assumptions, and has its advantages and limitations.

As shown in Chapter 3, the amino-acid pair predictability is subject to mutation. The great advantage in our methods is that we can numerically record the amino-acid pairs related to mutations.

In Section 3.2, we have analyzed several mutations in human Cx32 protein, and conducted detailed calculations to show how a mutation affects the amino-acid pairs, where house the mutation.

4.1.1. Amino-Acid Pair Targeted by Mutation

At first, we need to search the clue that which type of amino-acid pair the mutations target on. By answering this question, we can know which type of amino-acid pair sensitive to mutation.

The human Cx32 has recorded 188 missense point mutations, which can be classified according to the attribute of amino-acid pair targeted. We can count how many mutations occurred at predictable amino-acid pairs as well as unpredictable ones, by which we can have a concept that which types of amino-acid pairs are sensitive to mutation.

Table 4-1 shows our counting, we can see that 86.17% of mutations occur at unpredictable amino-acid pairs and only 13.83% of mutations occur at predictable amino-acid pairs. Thus, the unpredictable amino-acid pairs are more sensitive to mutation. This is very suggestive, because the results not only support our logic that randomness engineers mutation but also demonstrate the advantage of our methods. Although the predictable and unpredictable amino-acid pairs have nothing relevant to physicochemical property, an unpredictable amino-acid pair can become a predictable one after mutation or vice versa.

Furthermore, Table 4-2 lists the detailed classifications of amino-acid pairs targeted by mutations. 47.87% of mutations occurred at two amino-acid pairs, whose actual frequency is larger than their predicted one, and 31.38% of mutations occurred at the amino-acid pairs, of which one pair has a larger actual frequency than its predicted one, and the rest has no difference between its actual and predicted frequency.

In fact, what happened in real-life is that the mutation tried to narrow the difference between actual and predicted frequency, especially tried to reduce the actual frequency shown in Table 4-2. In other words, a protein is more likely to eliminate the amino-acid pairs whose actual frequency is larger than their predicted one, as if the protein feels them "uncomfortable".

We have summarized which types of amino-acid pairs are more likely to be targeted by mutation. This is fine, however one might wonder how we can determine these susceptible amino-acid pairs in a protein? Practically, this is quite simple, because we have already assigned each amino-acid pair in a protein with its actual and predictable frequency, and we can easily find these susceptible amino-acid pairs in a protein.

Table 4-1. Occurrences of mutations with respect to randomly predictable and unpredictable amino-acid pairs in human Cx32 protein

Amino-acid pair	Types		Pairs		Mutations		Ratio	
	No.	%	No.	%	No.	%	Mutations/type	Mutations/pair
Predictable	84	44.44	95	33.69	26	13.83	26/84=0.31	26/95=0.27
Unpredictable	105	55.56	187	66.31	162	86.17	162/105=1.54	162/187=0.87
Total	189	100	282	100	188	100	188/189=0.99	188/282=0.67

Table 4-2. Amino-acid pairs targeted by mutations in human Cx32 protein

Amino-acid pair	Pair I	Pair II	Mutations		Total
			No.	%	%
Predictable	AF = PF	AF = PF	26	13.83	13.83
Unpredictable	AF > PF	AF > PF	90	47.87	86.17
	AF > PF	AF = PF	59	31.38	
	AF > PF	AF < PF	6	3.19	
	AF < PF	AF = PF	7	3.72	
	AF < PF	AF < PF	0	0	

AF, actual frequency; PF, predicted frequency.

4.1.2. Amino-Acid Pair Appeared through Mutation

In the above section, we have discussed the amino-acid pairs targeted by mutation. Equally interesting are the amino-acid pairs that will appear through mutation because the targeted amino-acid pairs disappeared after mutation. Anyway, the amino-acid pairs that will appear would be either necessary or economically easy to find for protein construction. Thus, we should set a time point to analyze these amino-acid pairs. In this section we focus on the time point before mutation, which implies why a mutant protein needs certain types of amino-acid pairs.

Table 4-3 shows the amino-acid pairs appearing through mutation. As the time point is before mutation, 71.28% of mutations bring out one or both substituting amino-acid pairs that are absent in normal human Cx32 protein (AF = 0). Also, 55.31% of mutations result in one or both substituting amino-acid pairs with their actual frequency smaller than the predicted one. Likely, the protein favors the appearance of amino-acid pairs, whose appearance has been predicted, but does not appear.

Here, we can couple the effect of mutation with substituted and substituting amino-acid pairs. For substituted amino-acid pairs, the mutation is likely to remove the amino-acid pairs that appear more than their predicted frequency. For substituting amino-acid pairs, the mutation is likely to add the amino-acid pair that is predicted to appear but does not appear.

Table 4-3. Amino-acid pairs appeared through mutations in human Cx32 protein

Amino-acid pairs		Mutations		Total
Pair I	Pair II	No.	%	%
AF = 0, PF > 0	AF = 0, PF > 0	11 *	5.85	71.28
AF = 0, PF > 0	AF = PF = 0	13 *	6.91	
AF = 0, PF > 0	AF = PF > 0	38 *	20.21	
AF = 0, PF > 0	AF < PF, AF ≠ 0	6 *	3.19	
AF = 0, PF > 0	AF > PF	26 *	13.83	
AF = PF = 0	AF = PF = 0	9	4.79	
AF = PF = 0	AF = PF > 0	19	10.11	
AF = PF = 0	AF < PF, AF ≠ 0	0 *	0	
AF = PF = 0	AF > PF	12	6.38	

Table 4-3. Continued

| Amino-acid pairs | | Mutations | | Total |
Pair I	Pair II	No.	%	%
AF < PF, AF ≠ 0	AF < PF, AF ≠ 0	0 *	0	28.72
AF < PF, AF ≠ 0	AF = PF > 0	6 *	3.19	
AF < PF, AF ≠ 0	AF > PF	4 *	2.13	
AF = PF > 0	AF = PF > 0	10	5.32	
AF > PF	AF > PF	5	2.66	
AF = PF > 0	AF > PF	29	15.43	

AF, actual frequency; PF, predicted frequency. * indicates the mutations which result in one or both substituting amino acid pairs with their actual frequency smaller than the predicted one (totally 55.31%).

The overall effects of 188 mutations in human Cx32 protein with respect to substituted and substituting amino-acid pairs can be plotted as Figure 4-1. This figure can be read as follows. The y-axis simply records how many mutations occurred and the x-axis is the sum of difference between actual and predicted frequency calculated from both amino-acid pairs affected by mutations, i.e. $\Sigma(AF - PF)$. For instance, a mutation at position 11 substitutes 'S' for 'G' which results in two amino acid pairs 'LS' and 'SG' changing to 'LG' and 'GG', because the amino acid is 'L' at position 10 and 'G' at position 12. The actual and predicted frequencies are 4 and 3 for 'LS', 4 and 1 for 'SG', 0 and 2 for 'LG', and 0 and 1 for 'GG', respectively. Thus, the difference between actual and predicted frequency is 4 with regard to the substituted amino-acid pairs [(4 – 3) + (4 – 1)], and –3 to the substituting amino-acid pairs [(0 – 2) + (1 – 1)]. By this way, we can compare the frequency difference in the amino acid pairs affected by mutations. From Figure 4-1, we can see that the difference between actual and predicted frequency is 2.01±1.58 (mean ± SD, ranging from –1 to 6) for substituted amino-acid pairs. This means that the mutations occur in the amino-acid pairs that appear more than their predicted frequency. Meanwhile, the difference between actual and predicted frequency is –0.07±1.36 (ranging from –4 to 5) for substituting amino-acid pairs, which implies that the substituting amino-acid pairs are more randomly constructed, as their actual and predicted frequencies are about the same. Striking statistical difference is found between the substituted and substituting amino-acid pairs ($P < 0.0001$).

The difference between actual and predicted frequency represents a measure of randomness for constructing amino-acid pairs, i.e. the smaller the difference, the more random the construction of amino-acid pairs. In particular, (i) the larger the positive difference is, the more the randomly unpredictable amino-acid pairs are present; and (ii) the larger the negative difference is, the more the randomly unpredictable amino-acid pairs are absent [29-47, 63, 64, 68-73].

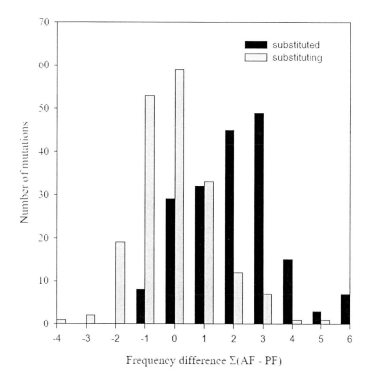

Figure 4-1. Overall effects of 188 mutations in human Cx32 protein for substituted and substituting amino-acid pairs.

4.2. Application of Fast Fourier Transform

In Section 3.3, we observed the behavior of predictable portion of hemagglutinins of influenza A viruses isolated from 1918 to 2006. Although the regression lines in Figure 3-5 indicate the general trend, we can still see the fluctuation of predictable portion.

These fluctuations are important because they are mutations in nature, otherwise there would be no changes in the predictable portions over time. We could be able to predict the time of mutation if we could find some pattern in Figure 3-5 because it presents the mutation history along the time course. However, it is obvious that we have little sense on the pattern in Figure 3.5.

To increase the possibility of pattern recognition, we have two ways, say, (i) we can increase the dataset along the time course, which however may need the data collected over several hundreds of years, and (ii) we can use mathematical tools to transfer the data in Figure 3-5 into other domain, where we can recognize its pattern. Likely, the second way is more promising because more data may still not enable us to recognize the pattern.

The mathematical tool we use to recognize the pattern over time is the fast Fourier transform (FFT), one of whose functions is to find the periodicity in chaotic data [81]. The best example, which serves us for understanding fast Fourier transform, is to find out the solar cycle using recorded sunspot data.

It is very important to find the periodicity for pattern recognition, because currently we do not know where a pattern begins and where a pattern ends. With periodicity, we can approximately determine the beginning and ending of pattern.

4.2.1. Periodicity of Predictable Portion along the Time Course

The determination of periodicity of predictable portion along the time course is in fact an easy job when using the fast Fourier transform. We generally accomplish this job using the MatLab [81].

Table 4-4 shows a dataset containing 2495 hemagglutinins of influenza A viruses, and the mean values of predictable portions for the given year are used for determination of periodicity. An issue that needs our attention is the data continuation along the time course, and we can see that the data continue from 1963 to present in Table 4-4. Thus, we can only use the data from 1963 to determine the periodicity. For MatLab, we simply input the mean values from 1963 to 2006, then use the standard steps to get the periodicity.

Figure 4-2 shows the periodicity of predictable portions based on the data from 1963 to 2006. Figure 4-2 can be read as follows. The unit of x-axis is years per cycle, which is the periodicity, thus each stick standing on x-axis represents a periodicity, and the more the sticks, the more the periodicities. The y-axis is the predictable portion of each periodicity, so the higher the predictable portion, the clear the periodicity. Accordingly, Figure 4-2 demonstrates several features.

1. As there are more than ten sticks in each panel, the predictable portion of amino-acid pairs along the time course contains many periodicities. In other words, the evolutionary process of influenza A virus hemagglutinins from 1963 to 2006 contains many periodicities.
2. As the periodicity of the predictable portion is different each other, each periodicity would contribute differently to the predictable portion. In other words, each periodicity suggests different number of mutations along the time course, i.e. the occurrence of mutations does not homogenously distribute from 1963 to 2006.
3. The periodicity with the biggest number of mutations is about 7.33 years in the upper and middle panels.

Table 4-4. Predictable portions of 2495 hemagglutinins used in Figure 3-5

Year	N	I	II	III	Year	N	I	II	III
1918	1	31.25	33.84	23.19	1974	9	36.33	40.53	28.47
1927	2	36.88	37.99	26.87	1975	10	35.50	39.84	28.28
1930	2	32.88	35.39	25.31	1976	28	36.60	40.89	28.55
1931	1	32.50	35.34	24.43	1977	45	36.43	40.46	28.46
1933	6	33.00	36.66	25.40	1978	18	35.26	39.44	28.18
1934	6	34.13	37.08	25.63	1979	28	38.19	42.52	30.57
1935	1	31.25	35.14	24.96	1980	27	36.69	40.55	28.29

Year	N	I	II	III	Year	N	I	II	III
1939	1	30.75	34.34	24.82	1981	11	36.21	41.31	29.39
1940	1	31.50	34.35	23.72	1982	14	35.77	39.12	27.01
1942	1	32.25	35.74	25.13	1983	49	35.72	39.46	28.97
1943	1	32.00	35.12	24.96	1984	32	35.23	41.19	30.93
1946	1	29.75	30.68	21.42	1985	33	37.74	42.23	29.62
1947	2	33.25	35.75	25.66	1986	20	36.95	41.23	29.33
1949	2	39.75	46.65	32.77	1987	20	38.26	42.83	30.60
1950	1	32.50	36.84	25.84	1988	23	36.82	41.10	29.36
1954	2	33.88	36.98	26.46	1989	11	37.82	42.18	30.16
1956	7	39.36	36.98	26.46	1990	8	36.88	40.39	28.95
1957	13	32.96	41.39	30.86	1991	20	36.59	40.92	28.91
1958	6	32.13	40.23	29.88	1992	14	36.71	40.63	28.90
1959	2	33.38	38.42	28.85	1993	16	36.55	40.78	28.85
1960	1	32.50	39.67	30.69	1994	29	37.76	41.60	29.85
1961	3	36.00	39.26	30.30	1995	49	34.70	37.57	26.23
1963	9	36.67	41.64	29.04	1996	73	35.30	38.98	27.79
1964	3	37.33	41.96	31.18	1997	87	34.56	39.37	28.52
1965	3	36.75	44.37	33.59	1998	93	35.86	40.31	29.06
1966	7	36.54	41.53	30.38	1999	178	36.52	41.16	30.01
1967	2	33.50	36.98	26.02	2000	201	36.25	40.73	29.32
1968	15	35.13	39.34	27.27	2001	185	34.61	38.74	27.56
1969	2	34.63	39.50	26.90	2002	206	34.47	39.25	28.44
1970	5	34.10	38.60	25.56	2003	265	35.76	40.07	29.06
1971	13	36.67	40.73	27.90	2004	279	35.32	39.66	29.31
1972	20	36.43	41.19	28.76	2005	218	34.92	39.18	28.41
1973	9	38.25	42.77	30.99	2006	45	34.21	38.74	27.67

N is the number of hemagglutinins in the given year. I, II and III are the predictable portions (%) with respect to 400 possible types, present types and frequencies, respectively.

Figure 4-2 can help us determine the periodicity, and then to determine the pattern of mutations in the hemagglutinins along the time course. Nevertheless, we would get a more precise picture if we would include more data. However, we only have the continuous data from 1963 to 2006, if we want to increase the size of dataset, thus we need to fill the empty cell in Table 4-4, such as the data in 1962. The common way to fill this empty cell is to calculate the mean based the data in 1961 and 1963, for our situation this mean is equal to 36.34% [(36 + 36.67)/2]. Now we can input the data from 1956 to 2006 to MatLab to calculate the periodicity.

Figure 4-3 shows the periodicity of predictable portions based on the data from 1956 to 2006. Comparing Figure 4-3 with Figure 4-2, we can see some difference in periodicity between two datasets, however the difference between two middle panels is insignificant.

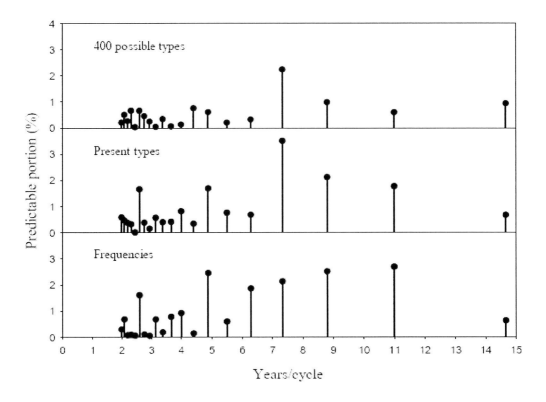

Figure 4-2. Periodicity of predictable portions of amino-acid pairs from influenza A virus hemagglutinins based on the data from 1963 to 2006

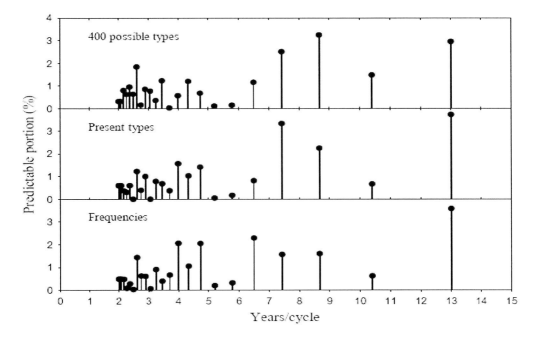

Figure 4-3 Periodicity of predictable portions of amino-acid pairs from influenza A virus hemagglutinins based on the data from 1956 to 2006

Thereafter, we can stratify the predictable portion along the time course in Figure 3-5 according to the determined periodicity in order to find the pattern of mutations.

4.2.2. Stratification of Predictable Portion According to Periodicity

The stratification of predictable portions of amino-acid pairs along the time course means that we divide the curves in Figure 3-5 into several parts, each part contains the number of years equal to the periodicity found by fast Fourier transform. There are many periodicities determined by fast Fourier transform, however, it is quite obvious that we can begin our stratification based on the periodicity with maximal predictable portion, which is about 7.33 years.

Here, we need to make approximation, because the hemagglutinins are documented in unit of year, so we cannot make the periodicity in fraction of year. Besides, the periodicity of 11 years is the biggest one in the lower panel in Figure 4-2. Clearly, this needs a number of trials. In this book, we use the periodicity of 9 years (average between 7 and 11) to stratify the predictable portion along the time course.

Another point in stratification is the beginning point for periodicity. As we consider the periodicity based on Figure 4-2, which begins from 1963, so the starting point for our stratification is 1963.

Figure 4-4 illustrates the stratification of predictable portions into 9-year periodicity. With the data from 1963 to 2006, we can group 5 cycles to analyze the possible pattern. Practically, one may not feel the pattern in Figure 4-4 clear enough, but this is the common situation we meet in real-life research, that is, the nature pattern is always not clear because there are so many factors involved. The pattern in Figure 4-4 looks like going up for 2 years, then going down for 2 years, then going up slightly for a year, and finally going somewhat stable but fluctuating for the rest 4 years.

In this figure, the hollow cycles indicate historical influenza epidemic, pandemic, or pandemic scares in humans. It is intriguing that there are three pairs of cycles on the same time points stratified by this periodicity.

An important application of such stratification is to estimate our current position during an evolutionary cycle [64, 82, 83]. For example, our position in 2006 is at the end of a cycle. Thus, we would expect that there would be more mutations in 2007 and 2008 because we are in the process of going up for the first 2 years of a cycle.

4.2.3. Searching Mutation Cause by Similar Periodicity in Nature

When looking at Figures 4-2 and 4-3, there are many periodicities determined by the fast Fourier transform, of which we have used only one or two for stratification of predictable portion along the time course.

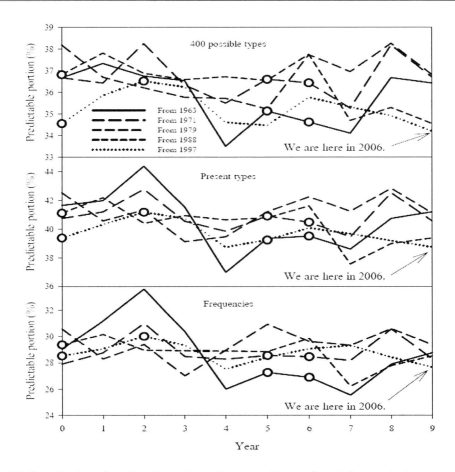

Figure 4-4. Stratification of predictable portions of amino-acid pairs from influenza A virus hemagglutinins along the time course. The hollowed cycles indicate influenza pandemic, epidemic, or pandemic scares.

Practically, these periodicities not only provide us with the possibility of stratification, by which we can estimate our current position along the hemagglutinin evolutionary process, but equally important provide us a way to find the mutation causes. This is also the application of obtained periodicity because we know that many nature phenomena have their own periodicity. For example, the periodicity that the Earth goes around its axis is about 24 hours.

On the other hand, we would expect there are many mutation causes if each periodicity in Figures 4-2 and 4-3 presents a nature cause, whose periodicity is similar to the periodicity in predictable portion of amino-acid pairs determined by fast Fourier transform.

A well-known periodicity is the sunspots, which has 10.8-year periodicity and somewhat equal to a periodicity around 11-year in Figure 4-2. Of course, this may be a coincidence between two nature phenomena. Anyway it would be worthy our attention, Figure 4-5 shows the periodicity of sunspot and neutron rate, comparing with the periodicity in predictable portions of amino-acid pairs in Figure 4-2. The coincidence may suggest that the cosmic rays play a certain role in the evolutionary process of influenza A virus hemagglutinins [84, 85].

Figure 4-6 shows the change in predictable portions of amino-acid pairs with respect to cosmic rays over time. In this figure, it is very hard for us to find any correlation between the mutation patterns and the cosmic rays over time. However, with the help of fast Fourier transform, we can find out several periodicities corresponded between the mutation patterns and cosmic rays. Once again, this is the great advantage of our method, say, we cannot find such interrelationship without applying the fast Fourier transform, we cannot apply the fast Fourier transform without quantification of proteins, and we cannot quantify the protein without our method.

In fact, there are several other periodicities, which require us to find out the event with similar periodicity in nature.

4.2.4 Type mutation and frequency mutation

The fast Fourier transform provides us a way to determine the periodicity of predictable portion of amino-acid pair along the time course, by which we know that the evolutionary process of influenza A virus hemagglutinins contains many different periodicities. This is the analysis along the time axis, but still another important question is how many mutations occurred in each periodicity [64, 82].

Actually, the height of sticks standing on x-axis in Figures 4-2, 4-3 and 4-5 is proportional to the number of mutations, so we have the concepts of type and frequency mutations to calculate how many mutations occurred in each periodicity.

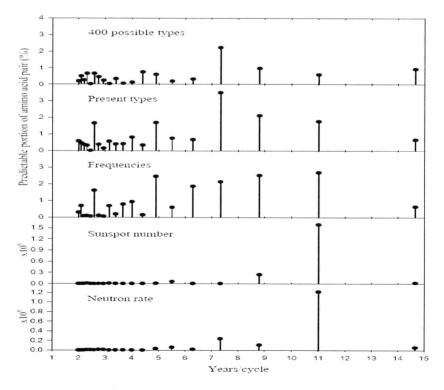

Figure 4-5. Periodicity of predictable portions of amino-acid pairs from influenza A virus hemagglutinins, sunspot number and neutron rate based on the data from 1963 to 2006.

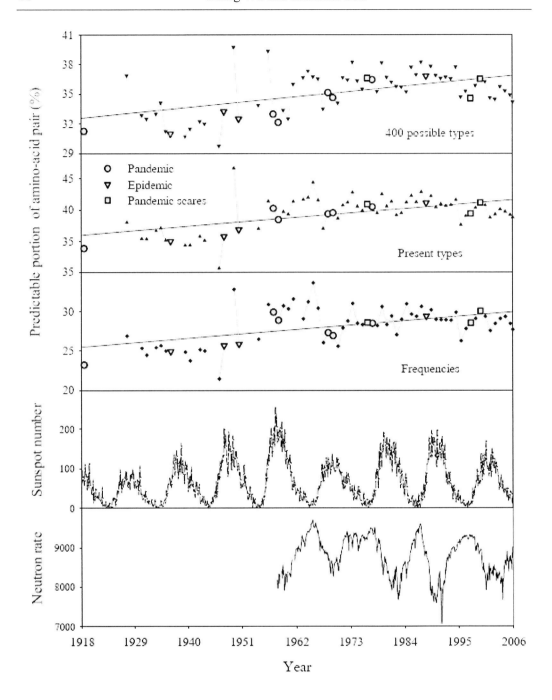

Figure 4-6. Change in predictable portions of amino-acid pair (the black symbols), sunspot number and neutron rate over time. The black lines are regressed lines for given predictable portion.

There are 400 types of theoretically possible amino-acid pairs and we use the 100% to classify them as predictable and unpredictable types, thus 0.25% represents one of 400 types, so a 0.25% change indicates that one of 400 types mutates to an unpredictable type from a predictable one or vice versa. This is the type mutation with respect to 400 possible types of amino-acid pairs.

Another type mutation is related to the present types of amino-acid pairs in a given protein. For example, the 1918 hemagglutinin (accession number AAD17229) has 263 present types of amino-acid pairs. Thus, about 0.38% change can be referred to the change in a present type.

Similarly, the frequency mutation is referred to the number of amino-acid pairs present in a protein. Again, the 1918 hemagglutinin contains 566 amino acids, which construct 565 amino-acid pairs. Hence, about 0.18% (1/565) change can be regarded as a modification in an amino acid pair in the hemagglutinin.

Now let us calculate the type and frequency mutations with respect to the periodicity in Figure 4-2. In the upper panel, the periodicity located at 7.33 years has the biggest predictable portion of 2.23%. As this panel related to 400 possible types, so we have 400 × 2.23% = 8.92 type mutations, that is, about 9 different types of amino-acid pairs mutated each 7.33 years. On the other hand, we can say that there will be about 9 mutations in different types of amino-acid pairs in future. Please note these 9 different types also include the absent types of amino-acid pairs.

In the middle panel, the periodicity located at 7.33 years has the biggest predictable portion of 3.50%. As this panel related to the present types of amino-acid pairs, which is about 271.31 for 2495 hemagglutinins, so we have 271.31 × 3.50% = 9.50 type mutations, that is, about 10 different present types of amino-acid pairs mutate each 7.33 years.

In the lower panel, the periodicity located at 7.33 years has the predictable portion of 2.13%. As this panel related to the present frequency of amino-acid pairs, which is about 565.34 for 2495 hemagglutinins, so we have 565.34 × 2.13% = 12.04 frequency mutations, that is, about 12 different present amino-acid pairs mutate each 7.33 years.

4.3. Timing of Mutation

The prediction of when a mutation will occur is a very important issue. However, we have mentioned in Chapter 1 that this prediction is very difficult although it is only related to two things, time and mutation. This is so because a mutation can be triggered by various known and unknown causes, thus the timing of mutation is almost impossible even if we would have determined all the causes. Besides, it is very difficult to determine whether a mutation cause is still effective, as the current version of protein might no longer be subject to the previous causes because of evolution.

On the other hand, we can approximately predict the time of occurrence of mutations with our methods, because they reliably and numerically record the history of the protein of interests as we have seen in Figure 3-5. Each line in Figure 3-5 in fact includes all the documented mutations occurred in history no matter whether or not the mutation causes were found.

We can get some sense on timing mutation if we study this fluctuating curve in Figure 3-5 in great details, naturally the more documented mutations are, the more correct the prediction is.

One way, for example, is to use the fast Fourier transform to determine the periodicity of the curve in Figure 3-5, which we have seen in this Chapter as shown in Figures 4-2 and 4-3.

The determination of periodicity of the curve in Figure 3-5 actually suggests that we have determined the periodicity of mutations because our methods have accurately and reliably recorded the evolutionary process. Thereafter we compare the pattern of each periodicity to determine our position in the current periodicity, and approximately predict the time of mutations, which is shown in Figure 4-4.

This is our current way for timing mutations in a large scale with respect to the calendar year [82, 83]. Of course, the timing of mutation would be more precise if the documentation of isolated proteins from influenza A virus would be conducted at smaller units such as month, week, and day. On the other hand, this means that at the best we can only make the prediction on when the wild mutation will occur at the same time unit as the documentation, now this unit is year.

Still, we can predict the number of mutations in each periodicity using type and frequency mutations described in Section 4.2.4.

4.4. Drug Target

One of interesting research fields over recent years is to use the bioinformatic tools to find the target for drugs [86], which sometimes is referred as the structure-based drug design. In general, a drug target can be the epitope and bounding sites in proteins [87, 88]. There are a number of web-based programs available for finding the targets, for example, the PROFsec method is used for protein secondary structure prediction [89-91], the average relative binding matrix method is used for T-cell epitope prediction [92, 93], the BepiPred method is used for B-cell epitope prediction [94, 95], and so on [96].

However, it is necessary to have some knowledge on the primary structure of protein in order to find a potential drug target. Moreover, the findings in most of studies focus on the drug target in a single protein.

Actually, the primary structure of protein can also be used for searching the drug target because the primary structure is the base of the secondary structure. Without detailed knowledge of functional sites in the protein in question, a probabilistically simple approach for drug efficacy is to target the abundant amino acids in the protein, because the drug would have a great chance to interact with the protein if the collision between drugs and proteins is a random event [97, 98].

Still, an important issue in finding the drug target is how to deal with the problem that the targeted amino acids have a larger chance of mutating. In such a case, the developed drug would be short-lived.

Furthermore, we need to consider whether the drug target can be used for a protein family, because each type of protein may contain many similar proteins.

4.4.1. Amino-Acid Pairs in A Protein Family

The counting of frequency of amino-acid pairs provides us with a measure to determine the amino-acid pairs, which are more likely to be hit by drug. We choose the amino-acid pairs rather than a single amino acid as drug targets because a good signature pattern of a protein must be as short as possible, but the conserved sequence is not longer than four or five residues [28].

To find drug target for a particular protein, we need to count the frequency of amino-acid pairs only in the particular protein. However, for the proteins from influenza A viruses, the finding of drug target needs to count the frequency of amino-acid pairs in many proteins, for example, a family of hemagglutinins, because there are so many different hemagglutinins at large.

Table 4-5 lists frequently appeared amino-acid pairs in hemagglutinins and neuraminidases, which are both surface proteins of influenza A viruses and the major antigens for neutralizing antibodies [99]. Table 4-5 can be read as follows, for example, we counted the frequency of all 400 possible types of amino-acid pairs in 131 hemagglutinins (column 2). Of counted frequency, the amino-acid pair "NG" is found in 121 hemagglutinins, which is 92.37% of 131. The similar reading can be applied to other columns in Table 4-5. The last column in Table 4-5 shows the amino-acid pairs in both hemagglutinins and neuraminidases.

In view of Table 4-5, we would expect that the drug has a large chance of missing its target if the targeted amino-acid pairs are not listed in Table 4-5, because these amino-acid pairs are the most frequently appeared ones, whereas the appeared frequency of other types of amino-acid pairs is smaller in all virus proteins, even much smaller than those in Table 4-5.

Comparing two surface proteins, the neuraminidases have more frequently appeared amino-acid pairs than hemagglutinins, so the neuraminidases are more suitable to serve as the drug target, which is consistent with other studies [100-102]. Nowadays the anti-neuraminidase drugs are wildly used in clinic, which conforms the advantage of our method that can throw lights on finding drug targets. Thus, the simple counting of the frequency of amino-acid pairs cross a protein family can provide us with a concept of universal target for the given protein family. Taking both hemagglutinins and neuraminidases into account, the frequency of appearance of amino-acid pairs in each hemagglutinin and neuraminidase is very small, only "NG", "TI" and "VK" can be found in more than 70% influenza A viruses.

No matter for a single protein or a protein family, the most frequently appeared amino-acid pairs would certainly have a large chance of hitting by drug. These amino-acid pairs can be easily found by counting the frequency of amino-acid pairs in a protein as well as in a protein family.

Table 4-5. Frequently appeared amino-acid pairs in hemagglutinins and neuraminidases from influenza A viruses

Amino-acid pair	Hemagglutinin (n=131)	Amino-acid pair	Neuraminidase (n=94)	Amino-acid pair	HA+NA (n = 225)
NG	121 (92.37%)	SG	94 (100.00%)	NG	189 (84.00%)
VK	109 (83.21%)	DN	93 (98.94%)	TI	181 (80.44%)

Table 4-5. Continued

Amino-acid pair	Hemagglutinin (n=131)	Amino-acid pair	Neuraminidase (n=94)	Amino-acid pair	HA+NA (n = 225)
GW	104 (79.39%)	EC	92 (97.87%)	VK	177 (78.67%)
FE	104 (79.39%)	SC	89 (94.68%)		
YH	102 (77.86%)	GS	82 (87.23%)		
AI	101 (77.10%)	TI	80 (85.11%)		
TI	101 (77.10%)	DG	76 (80.85%)		
CD	100 (76.34%)	RT	73 (77.66%)		
YP	100 (76.34%)	PN	71 (75.53%)		
ST	99 (75.57%)	NQ	70 (74.47%)		
SS	97 (74.05%)	RP	69 (73.40%)		
PK	95 (72.52%)	EL	69 (73.40%)		
SN	92 (70.23%)	NG	68 (72.34%)		
		PD	68 (72.34%)		
		SF	68 (72.34%)		
		TD	68 (72.34%)		
		VK	68 (72.34%)		
		IG	67 (71.28%)		
		LT	67 (71.28%)		
		WV	66 (70.21%)		

HA, hemagglutinin; NA, neuraminidase

4.4.2. Functional Amino-Acid Pairs

In searching drug target, one hopes the target to be the functional site in a protein. In this way, the efficacy of drug would be more profound. As we cannot clearly define all the functional sites in each protein, various approaches are developed to define the functional sites in a protein [103-106].

In view of amino-acid pair predictability, we would suggest that the amino-acid pair, whose difference between actual and predicted frequency is the largest, would be a functional site, because nature should deliberately make this difference by means of spending more energy and time.

Figure 4-7 shows several amino-acid pairs, which belong to the amino-acid pairs with larger difference between actual and predicted frequency among 329 H5N1 hemagglutinins, along the time course. Besides, these amino-acid pairs appear frequently in this database.

As can be seen, the difference of amino-acid pair "ME" is very stable along the time course. Actually, the amino-acid pairs shown in Figure 4-7 are rarely changed over time, in other words these amino-acid pairs are very conserved. Thus, we may consider them as the targets for development of drugs.

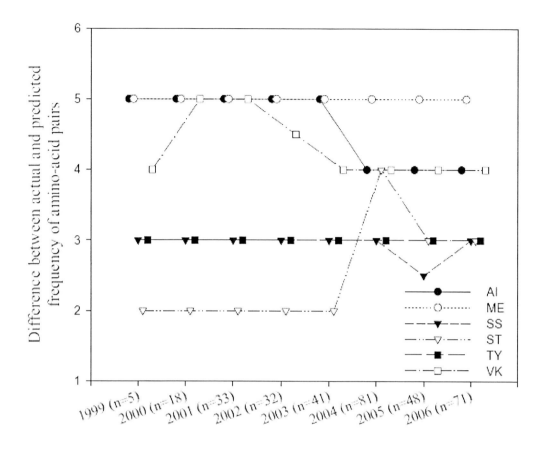

Figure 4-7. Difference between actual and predicted frequency in several amino-acid pairs over time. These data are obtained from the calculation of 329 H5N1 hemagglutinins from 1999 to 2006.

4.4.3. Amino-Acid Pairs with Less Chance of Mutating

At the earliest stage of developing amino-acid pair predictability, we also conducted several studies on calculating the first-order Markov chain probability [22, 24, 27, 107-113]. In plain word, the first-order Markov chain probability studies the probability that "e" follows "w", which constructs "we". Naturally, each amino acid has a certain probability to follow its preceding amino acid as "e" has a certain probability to follow "w" in English.

The implication here is that an amino acid would have a small chance of mutating if its first-order Markov chain probability is large, that is, an amino acid that has a large probability to follow its preceding amino acid is less sensitive to mutations than an amino acid that has a small probability to follow its preceding amino acid, because the larger the first-order Markov chain probability is, the more stable the amino-acid pair is [63].

Based on the three considerations from our approach, we propose the scheme of searching drug target in Figure 4-8, which presents an aspect in the drug discovery process. Actually we have already applied this scheme to searching the target for anti-SARS drugs in the structural proteins from SARS related coronavirus [38].

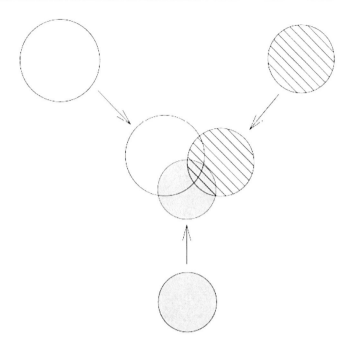

Figure 4-8. Drug targets (intersection among three circles) in relation to the amino-acid pairs (open circle) grouped according to their frequencies in proteins, the amino-acid pairs (lined circle) grouped according to the difference between their actual and predicted frequency and the amino-acid pairs (gray circle) grouped according to their first-order Markov chain probability.

Chapter 5

Amino-Acid Distribution Probability

Our first quantification, the amino-acid pair predictability, can provide us with a good measure on what probability an amino acid combines with another, although this measure does not provide the spatial position in a protein, that is, an amino-acid pair can principally locate anywhere in a protein without restriction.

5.1. Measurement of Spatial Randomness in Protein

Thus, we need to develop a measure to estimate the spatial position of an amino acid in a protein. Perhaps, the simplest way to answer this question is to make a guess according to the amino-acid composition in the protein in question. For example, there is one glutamine "Q" in human hemoglobin α-chain, meanwhile we know that human hemoglobin α-chain is composed of 142 amino acids, how can we guess the position of glutamine in this primary structure. The only answer is that this glutamine "Q" can be located at any position from 2 to 142 in this α-chain because the amino acid at the first position is methionine "M". This answer is nonsense, because it answers nothing. Likely, we do not find a rule for our guess that is random. In fact, this glutamine "Q" is located at position 54.

However, we can continue our guess because we have no other way to estimate the position of an amino acid without knowing other knowledge besides the composition of amino acids. Now let us make another guess using human hemoglobin β–chain, in which there are two methionines "M" among 147 amino acids. This time, we may generally imagine that we would divide this hemoglobin β-chain into two parts, each contains about 74 amino acids, then our guess is (i) both methionines are in the first half, (ii) one methionine is in each half and (iii) both methionines are in the second half. If we look at the real situation, we do find that both methionines are located in the first half, i.e. at positions 1 and 56.

Fortunately, we can find some rule in our guess because we have similar living experience. For instance, if we throw two balls into two holes, we can definitely know the possible results even before throwing, (i) two balls run into the first hole, (ii) two balls run into the second hole and (iii) each ball runs into each hole. This is very similar to our case. In fact, we can further simplify this case, that is, we do not need to distinguish the first and second holes. In this case, we have two possible results, (i) two balls run into a hole, and (ii) two balls run into both holes, which can correspond to our guess on how two methionines distribute along the hemoglobin β-chain, say, (i) two methionines are in each part and (ii) two methionines are in a part. Naturally, each case has the probability of 0.5 (Table 5-1).

Table 5-1. All possible distributions of two methionines "M" in human hemoglobin β-chain

Part 1	Part 2	Probability	Rank
M	M	0.5	1
MM		*0.5*	*1*

Bold and italic is the real distribution.

Although these results are not surprising, they do provide us with a probability, also throwing two balls into two holes is certainly a random event, because we can guess the general results but not the concrete result of each throwing. Not only we can guess the probability, but also we find that our problem is similar to the problem of occupancy of subpopulations and partitions [114]. In our case, we can calculate the distribution probability of a kind of amino acids along a protein sequence using the following equation:

$$\frac{r!}{r_1! \times r_2! \times ... r_n!} \times \frac{r!}{q_0! \times q_1! \times ... \times q_n!} \times n^{-r} \qquad \text{Equation 1}$$

where r is the number of amino acids, n is the number of divided parts and is equal to r in our case, r_n is the number of amino acids in the n-th part, q_n is the number of parts with the same number of amino acids, and ! is the factorial function, i.e. $n! = n \times (n-1) \times (n-2) \times ... \times 1$.

5.1.1. Distribution Probability of Two Amino Acids

Now let us see how to calculate the distribution probability of two methionines in Table 5-1. For the case that two methionines are in both parts (the second row in Table 5-1), we have $r = 2$ because we have two methionines, we have $n = 2$ as we imagine to divide the protein into two parts, $r_1 = 1$, $r_2 = 1$. Then we have $q_0 = 0$ because there is no part where methionine is not located, $q_1 = 2$ because there are two parts where a methionine is located, and $q_2 = 0$ because there is no part where two methionines are located. Thus we have the equation as $\frac{2!}{0! \times 2! \times 0!} \times \frac{2!}{1! \times 1!} \times 2^{-2} = \frac{2}{1 \times 2 \times 1} \times \frac{2}{1 \times 1} \times 0.25 = 0.5$ for this case.

Now let us look at the case that two methionines are in one part and the other part is empty (the third row in Table 5-1). The difference from the last case is $r_1 = 2$, $r_2 = 0$, $q_0 = 1$ because there is a part where no methionine can be found, $q_1 = 0$ because there is no part where a methionine is located, and $q_2 = 1$ because there is a part where two methionines can be found. So the equation is as follows $\dfrac{2!}{1!\times0!\times1!} \times \dfrac{2!}{2!\times0!} \times 2^{-2} = \dfrac{2}{1\times1\times1} \times \dfrac{2}{2\times1} \times 0.25 = 0.5$

Thus, it can be seen that there are two distribution patterns for two amino acids in a protein, and both probabilities are the same (0.5) so their distribution rank is 1 (Table 5-1).

5.1.2. Distribution Probability of Three Amino Acids

We can further attempt to guess the situations how three tyrosines "Y" distribute in human hemoglobin β-chain, which corresponds to throwing three balls into three holes. We imagine to dividing the human hemoglobin β-chain into three parts, and each part contains 49 amino acids (147/3). Table 5-2 shows all the possible distributions of three tyrosines. We can calculate their probabilities as we have done in the last section, and naturally we have $r = 3$, and $n = 3$.

For three tyrosines equally distributing in each part (the second row in Table 5-2), we have $r_1 = 1$, $r_2 = 1$, $r_3 = 1$, $q_0 = 0$, $q_1 = 3$, $q_2 = 0$, and $q_3 = 0$. Thus, we have

$$\frac{3!}{0!\times3!\times0!\times0!} \times \frac{3!}{1!\times1!\times1!} \times 3^{-3} = \frac{6}{1\times6\times1\times1} \times \frac{6}{1\times1\times1} \times \frac{1}{27} = 0.2222 .$$

For the distribution pattern in the third row of Table 5-2, we have $r_1 = 2$, $r_2 = 1$, $r_3 = 0$, $q_0 = 1$, $q_1 = 1$, $q_2 = 1$, and $q_3 = 0$, that is,

$$\frac{3!}{1!\times1!\times1!\times0!} \times \frac{3!}{2!\times1!\times0!} \times 3^{-3} = \frac{6}{1\times1\times1\times1} \times \frac{6}{2\times1\times1} \times \frac{1}{27} = 0.6667 .$$

For three tyrosines distributing in a single part (the last row in Table 5-2), we have $r_1 = 3$, $r_2 = 0$, $r_3 = 0$, $q_0 = 2$, $q_1 = 0$, $q_2 = 0$, and $q_3 = 1$, so

$$\frac{3!}{2!\times0!\times0!\times1!} \times \frac{3!}{3!\times0!\times0!} \times 3^{-3} = \frac{6}{2\times1\times1\times1} \times \frac{6}{6\times1\times1} \times \frac{1}{27} = 0.1111 .$$

Although there are three distribution patterns for three amino acids in a protein, their probabilities are different. The pattern listed in the third row of Table 5-2 has the highest probability (0.6667) and is ranked as the first.

Table 5-2. All possible distributions of three tyrosines "Y" in human hemoglobin β-chain

Part 1	Part 2	Part 3	Probability	Rank
Y	Y	Y	0.2222	2
YY	*Y*		*0.6667*	*1*
YYY			0.1111	3

Bold and italic is the real distribution.

Through the above two examples, it can be seen that we can easily calculate the distribution probability if we have a particular distribution pattern. Tables 5-3, 5-4 and 5-5 show the all distribution patterns for four, five and six amino acids, their distribution probabilities and ranks, respectively. Taking these tables into account, we can see several common trends:

(i)   The distribution probability decreases as the number of amino acids increases.
(ii) The distribution patterns do not increase proportionally to the increase in the number of amino acids. For example, two amino acids have 2 distribution patterns, three amino acids have 3 distribution patterns, four amino acids have 5 distribution patterns, five have 7, six have 11, seven have 15, and so on.
(iii) The probability of equal distribution, i.e. each part contains an amino acid, is quite small. Here, this point is somewhat different from our intuition on randomness, as we may usually consider that the randomness suggests the homogenous distribution. Actually, the homogenous distribution would be similar to the case that we would not expect to receive each letter per day if we would receive seven letters per week.

You might have noticed that we have introduced a new term, distribution rank, in these Tables, because (i) we cannot use the distribution probability for comparison between Tables because the distribution probability decreases as the number of amino acids increases, and (ii) there are the cases that the different patterns of distributions have the same distribution probability such as Table 5-1, and Table 5-5.

Table 5-3. All possible distributions of four asparagines "N" in human hemoglobin α-chain

Part 1	Part 2	Part 3	Part 4	Probability	Rank
N	N	N	N	0.0938	4
NN	*N*	*N*		*0.5625*	*1*
NN	NN			0.1406	3
NNN	N			0.1875	2
NNNN				0.0156	5

Bold and italic is the real distribution.

Table 5-4. All possible distributions of five serines "S" in human hemoglobin β-chain

Part 1	Part 2	Part 3	Part 4	Part 5	Probability	Rank
S	S	S	S	S	0.0384	5
SS	S	S	S		0.3840	1
SS	*SS*	*S*			*0.2880*	*2*
SSS	S	S			0.1920	3
SSS	SS				0.0640	4
SSSS	S				0.0320	6
SSSSS					1.6000e-3	7

Bold and italic is the real distribution.

Table 5-5. All possible distributions of six asparagines "N" in human hemoglobin β-chain

Part 1	Part 2	Part 3	Part 4	Part 5	Part 6	Probability	Rank
N	N	N	N	N	N	0.0154	5
NN	*N*	*N*	*N*	*N*		*0.2315*	*2*
NN	NN	N	N			0.3472	1
NN	NN	NN				0.0386	4
NNN	N	N	N			0.1543	3
NNN	NN	N				0.1543	3
NNN	NNN					6.4300e-3	7
NNNN	N	N				0.0386	4
NNNN	NN					9.6451e-3	6
NNNNN	N					3.8580e-3	8
NNNNN N						1.2860e-4	9

Bold and italic is the real distribution.

5.2. Comparison between Actual and Predicted Distribution Probability

Actually, a protein can only adopt one of distribution patterns for a kind of amino acids [23, 40, 68, 71]. For example, two methionines distribute in the first part of human hemoglobin β-chain, which is the actual distribution probability (see the bold and italic letters in Table 5-1). On the other hand, the randomness suggests that the event with the biggest probability is more likely to occur. Therefore, the maximal distribution probability can be viewed as the predicted distribution probability and can serve as reference for comparison. For instance, we can say that the amino acids methionines in Table 5-1, tyrosines in Table 5-2, and asparagines in Table 5-3 follow their predicted distribution probability, whereas the amino acids serines in Table 5-4 and asparagines in Table 5-5 do not follow their predicted distribution probability.

As we have shown in Chapter 3, the mutation has the tendency to minimize the difference between actual and predicted frequency of amino-acid pair. We may guess that the mutation has the tendency to minimize the difference between actual and predicted distribution probability or rank.

5.3. Predictable and Unpredictable Portions of Amino Acids

As have done in Chapter 2, where we classify the amino-acid pairs as predictable and unpredictable, we can also classify the amino acids as predictable and unpredictable according to their distribution probability. For example, the actual distribution probability of fifty amino acids in human hemoglobin α-chain equal to their predicted distribution probability, so the predictable portion of amino acids is 35.21% (50 amino acids/142 amino acids = 35.21%), while the unpredictable portion is 64.79% (100% - 35.21% = 64.79%) in the human hemoglobin α-chain. We therefore can use a single measure, either predictable portion or unpredictable one of amino acids to represent a whole protein with respect to its amino-acid distribution.

This way, we generally have two types of measures for comparison: we can use the predictable or unpredictable portion of amino acids if we would like to compare the difference among whole proteins, and we can also use the amino-acid distribution probability/rank for a kind of amino acids if we would like to compare the difference cross different kinds of amino acids.

5.4. Visualization of Distribution Probability in a Protein

With the same consideration in the visualization of amino-acid pair predictability in Chapter 2, we can assign the actual distribution probability (the top panel in Figure 5-1), predicted distribution probability (the middle panel in Figure 5-1), as well as the natural logarithm ratio of predicted distribution probability versus actual distribution probability "ln(predicted/actual)" (the bottom panel in Figure 5-1).

It is noticeable that we use the difference between actual and predicted frequency for visualization of the amino-acid pair predictability in Chapter 2. By contrast, here we use the natural logarithm ratio of predicted versus actual distribution probability for visualization. Although the use of logarithm is the very common practice in biomedical studies without reasoning, we have some strong argument for our case. In fact, whether or not we use logarithm is dependent on our aim, we will see that we need to make our measures in the similar scales in order to predict the mutation position in Chapters 10 and 11, otherwise each measure will have very different weight on the prediction model.

Similarly we can assign the actual distribution rank (the upper panel in Figure 5-2) and rank per amino acid (the lower panel in Figure 5-2) along the protein sequence

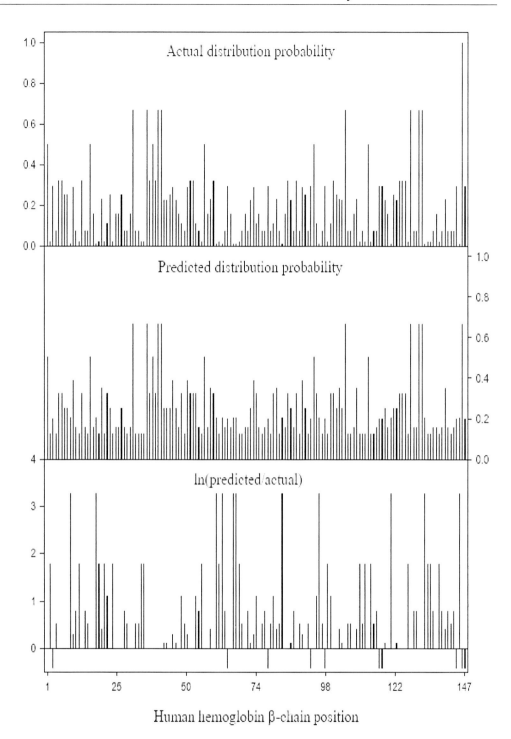

Figure 5-1. Visualization of amino-acid distribution probability in a protein.

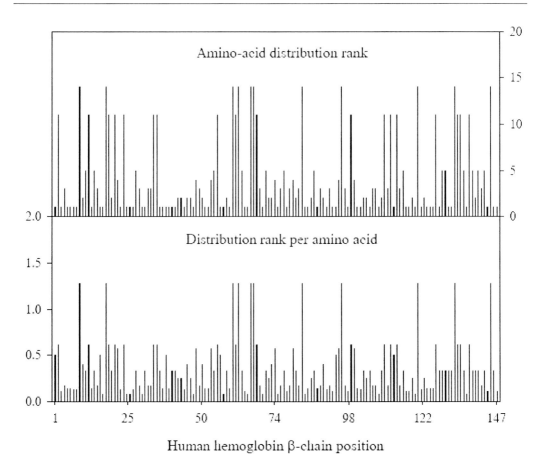

Figure 5-2. Visualization of amino-acid distribution rank in a protein.

Chapter 6

Behavior of Amino-Acid Distribution Probability

Similar to the detailed observations in Chapter 3, we need to observe the behavior of amino-acid distribution probability in various situations. Similarly, the behavior has two practical meanings, (i) we need to observe whether the amino-acid distribution probability is different cross proteins, cross subtypes, cross species, etc., and (ii) we need to observe whether the amino-acid distribution probability is different in a protein along the time course, such as before and after mutation.

Had we got the answer that the amino-acid distribution probability is no difference among various situations, we would conclude that the amino-acid distribution probability is a constant rather than a dynamic measure. We will follow the same path in Chapter 3 to treat this Chapter in the same frames.

6.1. Behavior of Amino-Acid Distribution Probability from Spatial Angle

6.1.1. Distribution Probability in Proteins with Different Lengths

Along the same line of thought, we observe the behavior of amino-acid distribution probability step-by-step. However, the careful reader might ask the following question for this section. The amino-acid distribution probability is different from the amino-acid pair predictability. Regarding to the amino-acid pair predictability, if the number of a kind of amino acids is the same for different lengths of proteins, not only the protein length will directly affect the calculation of permutation, but also the numbers of other kinds of amino acids will directly affect the calculation of permutation. However, the amino-acid distribution probability is only relevant to a single kind of amino acids in question, because we would group the protein according to the number of the kind of amino acids in question, how would it change if the proteins with different lengths have the same number of a single kind of amino acids?

Actually, whether the amino-acid distribution probability is the same depends on how the amino acids distribute rather than how we group the protein [63, 64]. In this view, the behavior of amino-acid distribution probability is in fact relevant to two issues, that is, (i) for the distribution probability of a kind of amino acids, we need to count if this kind of amino acids is the same cross proteins with different lengths, and (ii) for the distribution probability for a whole protein, we are almost certain that there is difference cross proteins with different lengths.

First, we investigate the behavior of amino-acid distribution probability with the same number of a certain kind of amino acids but belonging to proteins with different lengths. In Section 3.1, we have looked at eight proteins with different lengths and functions, so this time we can see these proteins again from the viewpoint of this Chapter.

Table 6-1 shows that only seven proteins have the amino acids with the same number. We can see that there are indeed the cases, where the number of amino acids is the same but the length of proteins is different in real-life. In fact it is quite surprising to see that both CA54 and FA9 have 32 asparagines "N" each, but their lengths are so different (1685 versus 461), for example.

Now let us conduct the exhausting computation on the distribution probability for these amino acids with the same number in Table 6-1.

Table 6-2 shows the detailed distributions for the amino acids with the same number in seven proteins. This table can be read as follows. From Table 6-1, we know CA54 and FA9 have 32 asparagines "N" each, and the distributions of "N" can be seen in Table 6-2. The first column in Table 6-2 indicates how we group the proteins according to the number of amino acids. For CA54 and FA9, each has 32 "N", so we can see that the columns "N" under CA54 and FA9 are filled with numbers.

Table 6-1. Amino acids with same number in seven proteins with different lengths

Protein	CA54	FA9	GLCM	HBA	LDLR	PH4H	VHL
Accession number	P29400	P00740	P04062	P01922	P01130	P00439	P40337
Amino acid	Number	Number	Number	Number	Number	Number	Number
R		20					20
N	32	32					
H	13	10	19	10	19	13	
M			11	3	11	3	3
F		27				27	
S				11			11
W						3	3
Length	1685	461	536	142	860	452	213

R, arginine; N, asparagine; H, histidine; M, methionine; F, phenylalanine; S, serine; W, tryptophan.

CA54, human collagen α5(IV) chain precursor; FA9, human coagulation factor IX precursor; GLCM, human glucosylceramidase precursor; HBA, human hemoglobin α-chain; LDLR, human low-density lipoprotein receptor precursor; PH4H, human phenylalanine-4-hydroxylase; VHL, human Von Hippel-Lindau disease tumor suppressor.

Table 6-2. Real-life distributions of amino acids with same number in Table 6-1

Part	CA54		FA9			GLCM			HBA			LDLR		PH4H				VHL			
	N	H	R	N	H	H	M	F	H	M	S	H	M	H	M	F	W	R	M	S	W
1	0	2	1	1	1	0	2	1	0	2	0	0	1	0	1	0	1	2	2	0	1
2	0	0	2	0	0	0	2	0	1	1	0	0	0	1	2	1	1	0	0	2	2
3	1	0	1	3	0	0	1	1	0	0	1	1	0	2	0	1	1	0	1	1	0
4	1	1	2	2	0	1	1	3	3		1	0	2	1		1		0		4	
5	2	0	0	0	0	0	0	0	1		1	2	0	2		2		0		0	
6	1	0	1	1	2	0	0	2	2		0	1	2	2		0		3		1	
7	1	1	1	2	4	2	1	0	1		2	1	1	1		0		1		1	
8	0	0	2	1	1	0	1	1	1		1	1	0	2		2		2		0	
9	0	1	0	2	2	1	1	1	1		0	2	3	2		1		0		1	
10	1	1	1	2	0	1	2	2	0		1	1	1	0		1		2		1	
11	0	0	0	1		3	0	0			3	1	1	0		0		2		0	
12	2	3	0	1		2		0				1		0		1		0			
13	2	4	2	0		2		2				1		0		2		0			
14	0		0	1		1		1				1				1		0			
15	0		1	2		1		3				2				2		1			
16	1		4	0		2		0				0				2		2			
17	0		1	1		1		0				2				0		2			
18	0		0	1		0		1				0				3		0			
19	1		1	1		2		2				2				0		2			
20	1		0	1				1								2		1			
21	1			4				0								1					
22	1			1				1								0					
23	2			1				2								1					
24	1			0				2								2					
25	4			0				0								1					
26	0			0				1								0					

Table 6-2. Continued

Part	CA54	FA9	GLCM	HBA	LDLR	PH4H	VHL
27	1	2	0			0	0
28	0	0					
29	6	0					
30	0	0					
31	2	1					
32	0	0					
I	$2.3936\text{e-}4$ 57	0.0301 0.0287	0.0286 0.0895 0.1616 0.0612	0.1524 0.6667 0.1077	0.0447 0.2020	0.0463 0.6667 0.0525 0.2222	$6.6980\text{e-}3$ 0.6667 0.0404 0.
II	145 9	10 8	8 2 2 2	2 1 3	7 1	6 1 3 2	24 1 8 1

I, Distribution probability; II, Distribution rank.

These numbers indicate whether we can find "N" in the grouped parts in CA 54 and FA9. For the column "N" under CA54, we do not find "N" in the first and second parts, but find one "N" in the third and fourth parts, respectively, and find two "Ns" in the fifth part, and so on. Meanwhile, we can see that the column "N" under FA9 has different distribution configuration from the column "N" under CA54. As asparagine "N" is the only kind of amino acids with the same number for CA54 and FA9, so we finished the comparison of "N" between CA54 and FA9. Following the same procedure, we can compare the distributions between other proteins.

Although we have compared the distribution of 32 "Ns" between CA54 and FA9, we may need to emphasize that each part contains very different number of amino acids, in CA54 there are 53 amino acids per part ($1685/32 = 52.66 \approx 53$), while there are only 14 amino acids per part in FA9 ($461/32 = 14.41 \approx 14$), that is, the amino-acid distribution probability does reflect the distribution pattern of amino acids. Likely, we would expect very few cases that the amino acids with the same number have the same distribution probability.

However, we can see the same distribution probability in the columns "M" under HBA and PH4H. Therefore, the distribution probability of amino acids with the same number can be the same or different case-by-case cross proteins with different lengths.

But it is impossible that two proteins with different length have the same numbers of all 20 kinds of amino acids, we would expect that the predictable and unpredictable portions of amino acids would be different cross these seven proteins with different lengths (Table 6-3).

Now let us check the predictable portion of amino acids according to their distribution probability in some large-scale (Table 6-4). Figure 6-1 shows the predictable portion of amino acids as well as the amino-acid distribution rank per amino acid based on 43 proteins listed in Table 6-4. In the lower panel of Figure 6-1, we can see that the predictable portion of amino acids is different cross proteins with different lengths. Besides, a trend observed in the lower panel is that the predictable portion decreases as the protein length increases.

In the upper panel, we show a derivate from amino-acid distribution probability. We calculate the distribution rank per amino acid in such a way, (i) each rank per a kind of amino acids is divided by the number of amino acids, for example, the rank in the column "N" under CA54 in Table 6-2 is 145, so we have 145/32 because there are 32 "Ns" in CA54; (ii) we calculate all 20 kinds of amino acids for a protein in this way, and (iii) we sum all divided ranks together. This derivate is useful for comparison cross proteins [117-120], because the amino-acid distribution probability would be too small if we calculate it in a similar way.

In this manner, we finish this section on observing behavior of amino-acid distribution probability and conclude that this measure is sensitive to protein length.

Table 6-3. Predictable and unpredictable portions of amino acids according to their distribution probability in five proteins

Protein	FA9	GLCM	HBA	PH4H	VHL
Predictable portion, %	3.25	0	35.21	17.04	11.74
Unpredictable portion, %	96.75	100	64.79	82.96	88.27

FA9, human coagulation factor IX precursor; GLCM, human glucosylceramidase precursor; HBA, human hemoglobin α-chain; PH4H, human phenylalanine-4-hydroxylase; VHL, human Von Hippel-Lindau disease tumor suppressor.

Table 6-4. Forty-three proteins with different lengths used in Figure 6-1

Protein	Accession no.	Source	Species	Length
Envelope protein	P59637	Corona virus	Human	76
Envelope protein	P24415	Corona virus	Porcine	82
Nonstructural protein	Q04854	Corona virus	Human	84
Matrix protein 2	Q9Q0L9	Influenza A virus	Goose	97
Matrix protein 2	Q9EAF2	Influenza A virus	Human	97
Nonstructural protein	Q04703	Corona virus	Canine	101
Nonstructural protein	Q04853	Corona virus	Human	109
Nonstructural protein 2	Q9Q0L7	Influenza A virus	Goose	121
Nonstructural protein 2	Q9DHF8	Influenza A virus	Human	121
Hemoglobin α-chain	P01922	Human	Human	142
Nonstructural protein	P33467	Corona virus	Feline	176
Von Hippel-Lindau disease tumor suppressor	P40337	Human	Human	213
Nonstructural protein	Q04704	Corona virus	Canine	213
Membrane protein	P59596	Corona virus	Human	221
Membrane protein	P03415	Corona virus	Murine	228
Nonstructural protein 1	Q9Q0L6	Influenza A virus	Goose	230
Nonstructural protein 1	Q9DHF7	Influenza A virus	Human	230
Matrix protein 1	Q9Q0L8	Influenza A virus	Goose	252
Matrix protein 1	Q91U69	Influenza A virus	Human	252
Nonstructural protein	P18517	Corona virus	Bovine	277
Replicase polyprotein 1ab	Q9WQ77	Corona virus	Rat	307
Pilx9 protein	Q9EUF2	*Escherichia coli*	*Escherichia coli*	309
Ornithine carbamoyltransferase	P00481	Rat	Rat	354
Ornithine carbamoyltransferase	P00480	Human	Human	354
Ornithine carbamoyltransferase	P11725	Mouse	Mouse	354
Nucleocapsid protein	O12298	Feline	Feline	376
Nucleocapsid protein	Q04700	Corona virus	Feline	382
TP53	P04637	Human	Human	393
Nucleocapsid	P59595	Corona virus	Human	422
Hemagglutinin-esterase precursor	AY316300	Equine	Equine	423
Hemagglutinin-esterase precursor	P30215	Corona virus	Human	424
Nucleocapsid protein	P10527	Corona virus	Bovine	448
Phenylalanine-4-hydroxylase	P00439	Human	Human	452
Coagulation factor IX precursor	P00740	Human	Human	461
Neuraminidase	Q9ICY2	Influenza A virus	Human	467
Neuraminidase	Q9Q0U7	Influenza A virus	Goose	469
Nucleoprotein	Q9Q0U8	Influenza A virus	Goose	498
Nucleoprotein	Q9ICY7	Influenza A virus	Human	498
Glucosylceramidase precursor	P04062	Human	Human	536
Hemagglutinin	Q9ICY5	Influenza A virus	Human	560
Hemagglutinin	Q9Q0U6	Influenza A virus	Goose	568
Bruton's tyrosine kinase	Q06187	Human	Human	659
Bruton's tyrosine kinase	P35991	Mouse	Mouse	659

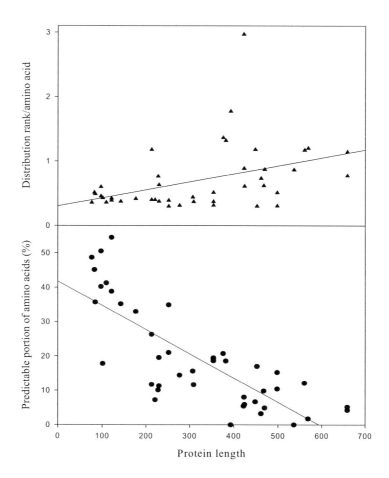

Figure 6-1. Predictable portion of amino acids and distribution rank per amino acid based on 43 proteins in Table 6-4.

6.1.2. Distribution Probability in Proteins with Similar Length

For proteins with similar length or the same length, we should at first look at if these proteins belong to the same function family. If so, it would be more efficient to look at the amino acids with different numbers because the amino acids with the same number are likely to distribute in the same manner. For example, Table 6-5 lists two H3N2 hemagglutinins with the same length isolated in 2006 from North America, which belong to the same function family. As seen in Table 6-5, these two hemagglutinins are very highly similar each other except for asparagine "N" and lysine "K", whose number and distribution probability indeed differ between two hemagglutinins.

Next, we look at the proteins with the same length but different function. For these proteins, we should at first compare if the distribution probability is different in the amino acids with the same numbers. In Table 6-6, we can see that both nonstructural protein from canine corona virus and human Von Hippel-Lindau disease tumor suppressor have the length of 213 amino acids, of which aspartic acid "D" and leucine "L" are the same-number-amino-acids. As can be seen in Table 6-6, their distribution probability is different one another.

Table 6-5. Amino-acid number and distribution probability in two H3N2 hemagglutinins

	ABG80436	Distribution probability	ABG88817	Distribution probability
A	31	0.0323	31	0.0323
R	27	0.0250	27	0.0250
N	48	0.0188	47*	0.0053
D	29	0.0128	29	0.0128
C	18	0.0077	18	0.0077
E	28	0.0001	28	0.0001
Q	24	0.0714	24	0.0714
G	42	0.0033	42	0.0033
H	11	0.0808	11	0.0808
I	49	0.0202	49	0.0202
L	44	0.0231	44	0.0231
K	35	0.0437	36*	0.0348
M	8	0.0673	8	0.0673
F	23	0.0460	23	0.0460
P	20	0.0208	20	0.0208
S	40	0.0004	40	0.0004
T	34	0.0285	34	0.0285
W	11	0.1077	11	0.1077
Y	19	0.0497	19	0.0497
V	25	0.0209	25	0.0209
Length	566		566	

*, amino acids different between two hemagglutinins.

Table 6-7 shows the real-life distribution of aspartic acid "D" and leucine "L" in these two proteins. Although the distribution probability of "D" is very similar for both proteins in Table 6-6 (0.1010 versus 0.1077), their real-life distributions are very different in Table 6-7. A point needed our attention is that the grouped part is exactly the same for both proteins because they have the same length, thus each part contains 19 amino acids (213/11 = 19.36 ≈ 19) for "D" and 11 amino acids (213/20 = 10.65 ≈ 11) for "L", respectively.

Table 6-6. Amino-acid number and distribution probability in two proteins with the same length but different function

Protein	Nonstructural protein from canine corona virus (Q04704)		Human Von Hippel-Lindau disease tumor suppressor (P40337)	
Amino acid	Number	Distribution probability	Number	Distribution probability
A	5	0.1920	10	0.0476
R	10	0.1905	20	6.6980e-3

Protein	Nonstructural protein from canine corona virus (Q04704)		Human Von Hippel-Lindau disease tumor suppressor (P40337)	
Amino acid	Number	Distribution probability	Number	Distribution probability
N	5	0.3840	9	0.1770
D	11*	0.1010	11*	0.1077
C	8	0.2243	2	0.5000
E	14	0.1649	30	1.3566e-4
Q	7	0.2142	8	0.0673
G	12	0.1163	18	0.0389
H	11	6.7330e-3	5	0.0640
I	13	0.1158	6	0.1543
L	20*	0.0804	20*	0.0422
K	18	0.0138	3	0.1111
M	3*	0.6667	3*	0.6667
F	12	0.0310	5	0.2880
P	9	0.1967	19	0.0319
S	16	0.0639	11	0.0404
T	9	0.0197	7	0.2142
W	2	0.5000	3	0.6667
Y	15	0.0392	6	0.2315
V	13	0.1544	17	0.1280
Total	213		213	

*, amino acids with the same number in both proteins.

Table 6-7. Real-life distribution of aspartic acid "D" and leucine "L" in nonstructural protein from canine corona virus and human Von Hippel-Lindau disease tumor suppressor

Part	Nonstructural protein from canine corona virus (Q04704)		Human Von Hippel-Lindau disease tumor suppressor (P40337)	
	D	L	D	L
I	0	3	1	0
II	1	1	1	0
III	0	0	0	0
IV	1	2	0	0
V	2	1	1	1
VI	2	3	0	1
VII	1	0	2	0
VIII	2	0	1	1
IX	0	1	1	1
X	2	1	3	1
XI	0	1	1	2
XII		1		2
XIII		1		2

Table 6-7. Continued.

Part	Nonstructural protein from canine corona virus (Q04704)	Human Von Hippel-Lindau disease tumor suppressor (P40337)
XIV	0	1
XV	2	2
XVI	0	1
XVII	0	2
XVIII	2	2
XIX	0	1
XX	1	0

Finally, Figure 6-2 illustrates the predictable portion of amino acids for the influenza A virus hemagglutinins presented in Table 3-5. In principle, this figure can be read exactly as the same as Figure 3-2. The x- and y-axes indicate the hemagglutinin length and predictable portion in different contexts. A filled cycle marks the mean of length of given hemagglutinins with reference to x-axis, and the mean of predictable portion with reference to y-axis. Similarly, a capped line along x-axis direction is the standard deviation (SD) of length of given hemagglutinins, and a capped line along y-axis direction is the standard deviation (SD) of predictable portion. As we use the same means and SDs for x-axis for each panel, we only mark the hemagglutinin subtypes in the lower panel.

However, a particular difference is that the predictable portions of amino acids in H8 and H12 are so near to zero that the lower part of their SD is below zero after the calculation.

Anyway, Figure 6-2 confirms the behavior of predictable portion of amino acids different in the proteins with similar length, even in those with the same function.

6.1.3. Distribution Probability in Various Subtypes of Proteins

Until this point, we are quite sure that the amino-acid distribution probability would be different among various subtypes of proteins, because the behavior of amino-acid distribution probability has passed the strictest test, that is, the same-number-amino-acids generally have different distribution probabilities.

Since the difference in amino-acid composition between proteins from different families would be bigger than that between subtypes, what is the meaning for us to observe the behavior of amino-acid distribution probability in different subtypes?

However, we consider such an observation important because the classification of subtypes is according to some experimental techniques. Without a particular technique, we could not distinguish a particular classification based on this technique. Thus, we may ask ourselves whether our method can distinguish the subtypes of proteins classified by other techniques. If so, we would be very happy because we do not need the specific technique but can make the distinction.

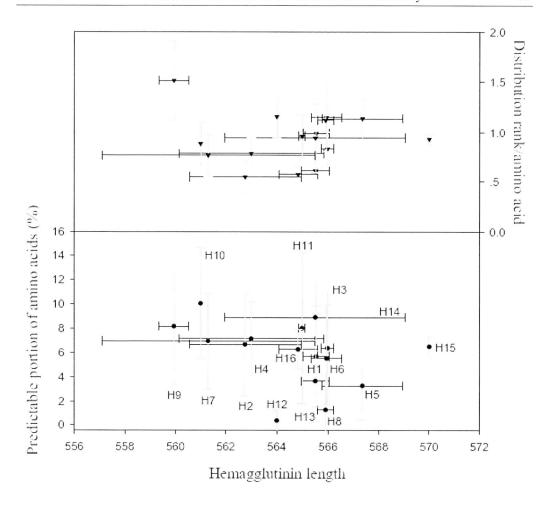

Figure 6-2. Predictable portion of amino acids and distribution rank per amino acid versus protein length in sixteen subtypes of influenza A virus hemagglutinins. The data are presented as mean±SD.

The best way, perhaps, to detect whether our method can distinguish subtypes of proteins is to choose two proteins from different subtypes with similar or at best the same length, and to see whether they are different in our sense. Figure 6-3 shows such an example, where two hemagglutinins have the same length of 566 amino acids, but are classified as different subtypes H1 and H3. Along x-axis of Figure 6-3 is the amino-acid composition. As can be seen, there are three kinds of amino acids with equal numbers between two hemagglutinins, methionine "M", proline "P" and asparagine "N", however their distribution ranks are different one another in y-axis. Furthermore, y-axis indicates the difference between two hemagglutinins, so that we can distinguish these two subtypes using our method.

Figure 6-4 shows the behavior of amino-acid distribution probability cross subtypes of 2495 hemagglutinins from influenza A viruses (Table 3-5). We can see that the amino-acid distribution probability successfully distinguishes the different subtypes. Equally, we can also say that the amino-acid distribution probability is different among subtypes of proteins [83].

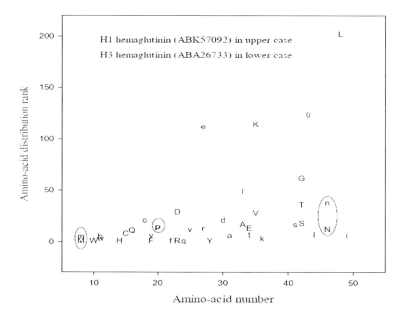

Figure 6-3. Amino-acid number and distribution rank in H1 hemagglutinin (strain A/Philippines/344/2004(H1N2)) and H3 hemagglutinin (strain A/Christchurch/14/2004(H3N2)). The ellipses indicate the kinds of amino acids have the same number in both hemagglutinins

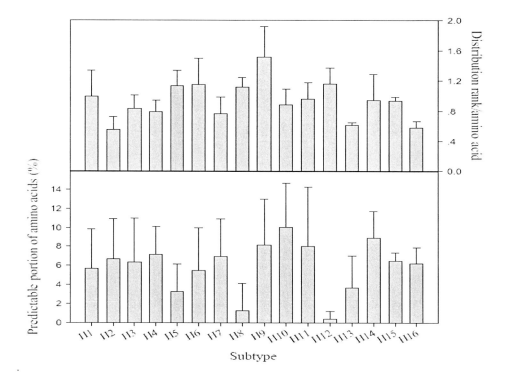

Figure 6-4. Predictable portion of amino acids and distribution rank per amino acid in different subtypes of influenza A virus hemagglutinins (the number of hemagglutinins in each subtypes can be found in Table 3-5). The data are presented as mean±SD.

6.1.4. Distribution Probability in Proteins Cross Species

Until now in this Chapter, we enlarge the scale of our observations from the proteins with the same length to proteins belonging to different subtypes, and now we reach the stage of observing the behavior of amino-acid distribution probability in proteins cross species in this section.

On the other hand, this process of observation is also the research process with the exponential increase in protein database, that is, we can conduct the research by comparing proteins cross subtypes and species to get more insight from random viewpoint [83].

Figure 6-5 shows the behavior of amino-acid distribution probability cross species, where we can see the difference among species. It is noticeable that these measures are quite similar in avian and human H5 hemagglutinins, suggesting that avian H5 viruses are easy to transform to humans.

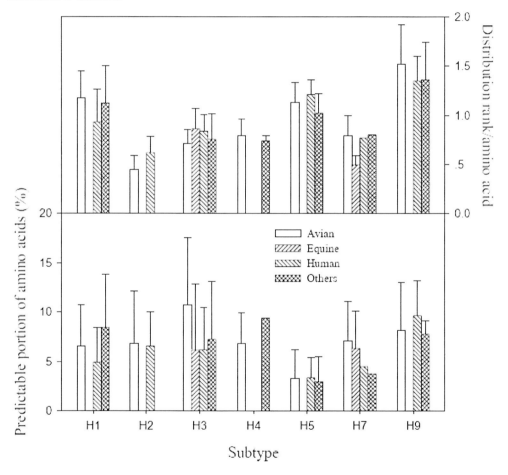

Figure 6-5. Predictable portion of amino acids and distribution rank per amino acid in hemagglutinins from influenza A viruses cross species (the number of hemagglutinins in each species and subtype can be found in Table 3-6). Data are presented as mean±SD.

6.2. Behavior of Amino-Acid Distribution Probability from Time Angle

In this section, we mainly observe the behavior of amino-acid distribution probability along the time course, i.e. before and after mutation. Without a mutation, we cannot find anything related to the change in primary structure of a protein, although the protein function can be different from time to time, which is more relevant to high-level structures of the protein [115, 116].

As nature is too complex to understand fully, the researchers have to stratify the problem into several levels in order to concentrate themselves on a niche, where they can be profitable. We stratify our observation into three levels, (i) whether a mutation changes the actual distribution probability of the kind of amino acids, of which one amino acid disappears after mutation, (ii) whether a mutation changes the actual distribution probability of the kind of amino acids, of which one amino acid appears after mutation, and (iii) whether a mutation changes the actual distribution probability of a whole protein.

Again, we take the human hemoglobin β-chain (accession number P02023) as an example. This hemoglobin is composed of 147 amino acids, and Table 6-8 shows its actual and predicted amino-acid distribution probability and rank, which actually is the first time in this book that we can see the details in a single protein. From this table, we can see several features:

(i)   More than a half of 20 kinds of amino acids have the actual distribution probability equal to their predicted one, and their distribution rank is marked as 1, indicating that these kinds of amino acids distribute in this protein with the maximal distribution probability.

(ii) The rest kinds of amino acids posses the actual distribution probability different from their predicted one, and the smaller the distribution probability is, the bigger the distribution rank is.

(iii) The actual distribution probability is far away from the equal distribution probability that each part contains an amino acid in question, thus we would not expect to see that amino acids would distribute with equal distance, that is, amino acids cannot homogenously distribute in a protein sequence.

Table 6-8. Actual and predicted distribution probability in human hemoglobin β-chain

Amino acid	Number	Actual distribution probability	Predicted distribution probability	Equal distribution probability	Rank
A	15	0.0706	0.1569	2.9863e-6	5
R	3	0.6667	0.6667	0.2222	1
N	6	0.2315	0.3472	0.0154	2
D	7	0.1071	0.3213	6.1199e-3	4
C	2	0.5000	0.5000	0.5000	1
E	8	0.2523	0.2523	2.4033e-3	1
Q	3	0.6667	0.6667	0.2222	1
G	13	0.1544	0.1544	2.0560e-5	1
H	9	0.1967	0.1967	9.3666e-4	1

Amino acid	Number	Actual distribution probability	Predicted distribution probability	Equal distribution probability	Rank
I	0	- -	- -	- -	- -
L	18	0.0748	0.1246	1.6272e-7	3
K	11	7.6948e-3	0.2020	1.3991e-4	14
M	2	0.5000	0.5000	0.5000	1
F	8	0.2243	0.2523	2.4033e-3	2
P	7	0.3213	0.3213	6.1199e-3	1
S	5	0.2880	0.3840	0.0384	2
T	7	0.3213	0.3213	6.1199e-3	1
W	2	0.5000	0.5000	0.5000	1
Y	3	0.6667	0.6667	0.2222	1
V	18	0.0208	0.1246	1.6272e-7	11

6.2.1. Distribution Probability Before Mutation

Now let us look at how the amino-acid distribution probability behaves at the three levels we stratified above. In human hemoglobin β-chain, a kind of frequently mutating amino acids is glutamic acids "E", which are located at positions 7, 8, 23, 27, 44, 91, 102 and 122 in human hemoglobin β-chain. Because there are eight glutamic acids, we imagine to grouping this human hemoglobin β-chain into 8 parts, of which each contains 18.375 amino acids as the human hemoglobin β-chain contains 147 amino acids. In our calculation, we use 18 amino acids for the first seven parts, and the last part contains 21 amino acids, i.e. 18 amino acids × 7 parts = 126 amino acids and 147 amino acids – 126 amino acids = 21 amino acids. Table 6-9 lists the distribution pattern of the eight glutamic acids.

Table 6-9. Actual distribution of eight glutamic acids "E" with respect to positions in human hemoglobin β-chain

Part	I	II	III	IV	V	VI	VII	VIII
Position	1-18	19-36	37-54	55-72	73-90	91-108	109-126	127-147
Distribution	EE	EE	E		E	E	E	

Now we can calculate the actual distribution probability of the distribution pattern shown in Table 6-9. According to the Equation 1 in Chapter 5, we have

$$\frac{r!}{r_1! \times r_2! \times ... r_n!} \times \frac{r!}{q_0! \times q_1! \times ... \times q_n!} \times n^{-r}$$

$$= \frac{8!}{2! \times 2! \times 1! \times 0! \times 1! \times 1! \times 1! \times 0!} \times \frac{8!}{2! \times 4! \times 2! \times 0! \times 0! \times 0! \times 0! \times 0! \times 0!} \times 8^{-8}$$

$$= \frac{40320}{2 \times 2 \times 1 \times 1 \times 1 \times 1 \times 1 \times 1} \times \frac{40320}{2 \times 24 \times 2 \times 1 \times 1 \times 1 \times 1 \times 1 \times 1} \times \frac{1}{16777216} = 0.2523$$

i.e. the actual distribution probability is 0.2523, which in fact corresponds to the predicted distribution, the maximal distribution probability (Table 2 in Appendix).

6.2.2. Distribution Probability of Substituted Amino Acids

The mutation in position 23 leads glutamic acid "E" to change to several other types of amino acids. Now let us do some calculations to answer the three questions related to the behavior of amino-acid distribution predictability from time angel.

Our first question is if a mutation changes the actual distribution probability of the kind of amino acids, of which one disappears after mutation.

What will happen for glutamic acids when the mutation occurs at position 23? Obviously, the number of glutamic acids will reduce to seven from eight, so their distribution probability must be calculated according to seven glutamic acids in the protein. Accordingly, we imagine to grouping the human hemoglobin β-chain into 7 parts, and each part contains 21 amino acids (147 amino acids/7 parts = 21 amino acids).

Table 6-10. Actual distribution of seven glutamic acids "E" with respect to positions in human hemoglobin β-chain after mutation at position 23

Part	I	II	III	IV	V	VI	VII
Position	1-21	22-42	43-63	64-84	85-105	106-126	127-147
Distribution	EE	E	E		EE	E	

According to Table 6-10 and Equation 1 in Chapter 5, we have

$$\frac{r!}{r_1! \times r_2! \times \ldots r_n!} \times \frac{r!}{q_0! \times q_1! \times \ldots \times q_n!} \times n^{-r}$$

$$= \frac{7!}{2! \times 1! \times 1! \times 0! \times 2! \times 1! \times 0!} \times \frac{7!}{2! \times 3! \times 2! \times 0! \times 0! \times 0! \times 0! \times 0!} \times 7^{-7}$$

$$= \frac{5040}{2 \times 1 \times 1 \times 1 \times 2 \times 1 \times 1} \times \frac{5040}{2 \times 6 \times 2 \times 1 \times 1 \times 1 \times 1 \times 1} \times \frac{1}{823543} = 0.3213$$

i.e. the actual distribution probability of seven glutamic acids is 0.3213, which is larger than that before mutation (0.2523). Thus, the actual distribution probability is different before and after mutation, and this example answers our question of whether a mutation changes the actual distribution probability of the kind of amino acids, of which one amino acid disappears after mutation.

6.2.3. Distribution Probability of Substituting Amino Acids

Now let us answer the second question that whether a mutation changes the actual distribution probability of the kind of amino acids, of which one amino acid appears after mutation. Here, we take a kind of amino acids, which are few in human hemoglobin β-chain, in order to facilitate the presentation of distribution.

There is a mutation that occurs at position 23 changing glutamic acid "E" to glutamine "Q". Before mutation there are three glutamines, whose positions are 40, 128 and 132, so we imagine to grouping the human hemoglobin β-chain into three parts, and each contains 49 amino acids (147 amino acids/3 parts = 49 amino acids).

Table 6-11. Actual distribution of three glutamines "Q" in human hemoglobin β-chain before mutation

Part	I	II	III
Position	1-49	50-98	99-147
Distribution	Q		QQ

Accordingly, the actual distribution probability in Table 6-11 is equal to

$$\frac{r!}{r_1!\times r_2!\times...r_n!}\times\frac{r!}{q_0!\times q_1!\times...\times q_n!}\times n^{-r}$$

$$=\frac{3!}{2!\times 1!\times 0!}\times\frac{3!}{1!\times 1!\times 1!\times 0!}\times 3^{-3}=\frac{6}{2\times 1\times 1}\times\frac{6}{1\times 1\times 1\times 1}\times\frac{1}{27}=0.6667$$

Actually, this distribution probability (0.6667) is equal to the predicted one. The mutation at position 23 changes glutamic acid to glutamine so that there are four glutamines after mutation. Similarly, we calculate the actual distribution as follows. We divide the human hemoglobin β-chain into 4 parts, and each part contains 36.75 amino acids (147 amino acids/4 parts = 36.75 amino acids). Thus in our calculation, we use 37 amino acids for the first three parts, and the last part contains 36 amino acids (37 amino acids × 3 parts = 111 amino acids, and 147 amino acids – 111 amino acids = 36 amino acids).

Table 6-12. Actual distribution of four glutamines "Q" in human hemoglobin β-chain after mutation

Part	I	II	III	IV
Position	1-36	37-72	73-108	109-147
Distribution	Q	Q		QQ

Again, we can calculate the distribution probability in Table 6-12 as follows.

$$\frac{r!}{r_1!\times r_2!\times...r_n!}\times\frac{r!}{q_0!\times q_1!\times...\times q_n!}\times n^{-r}$$

$$=\frac{4!}{1!\times 1!\times 0!\times 2!}\times\frac{4!}{1!\times 2!\times 1!\times 0\times 0!}\times 4^{-4}=\frac{24}{2\times 1\times 1\times 1}\times\frac{24}{1\times 2\times 1\times 1\times 1}\times\frac{1}{256}=0.5625$$

In fact, this actual distribution probability is equal to their predicted one. However, we do notice the actual distribution different before and after mutation, this way, we answer our second question of if a mutation changes the actual distribution probability of the kind of amino acids, of which one amino acid appears after mutation, and the answer is yes!

6.2.4. Predictable Portion before and after Mutation

Before answering the third question of if a mutation changes the actual distribution probability of a whole protein, it is necessary to know how many ways can be used to present a whole protein with respect to the actual distribution probability. We can simply sum all the actual distribution probabilities together, and we can also calculate the actual distribution probability per amino acid, which are common methods in biomedical settings. However, the simplest way perhaps is to use the predictable and unpredictable portions of amino acids.

We firstly look at whether the predictable and unpredictable portions change before and after mutation. The predictable portion in human hemoglobin β-chain before mutation can be calculated using Table 6-8, where the actual and predicted distribution probabilities are the same in 11 kinds of amino acids (R, C, E, Q, G, H, M, P, T, W, and Y). The total number of these amino acids is 59, so the predictable portion is equal to 40.14% (59 amino acids/147 amino acids = 40.14%).

In the case that glutamic acid "E" at position 23 changes to glutamine "Q", according to the new composition of 7 glutamic acids and 4 glutamines in the mutant hemoglobin, the predictable portion is 40.14%, which is the same as the one before mutation, so this mutation does not change the predictable portion of amino acids.

For another case, when glutamic acid "E" at position 23 is replaced by alanine "A", the predictable portion is 39.46%, which is different from the one before mutation, thus the mutation does change the predictable portion of amino acids.

Both cases indicate that a mutation either can or cannot change the predictable portion of amino acids in a protein, which answers our third question. The answer is somewhat against our simple deduction that the predictable portion would change because the actual distribution probability changes in original and mutated amino acids. This again demonstrates the difference between living and dead measures in protein, i.e. the mystery of nature makes the property of a whole protein equal or unequal to the sum of its components. This characteristic further supports our development of living measures in proteins.

6.3. Behavior of Amino-Acid Distribution Probability from Spatial and Time Angles

As we have done in Chapter 3, we now need to observe the behavior of amino-acid distribution probability from both spatial and time angles. With the dataset in Table 3-15, we can plot the amino-acid distribution probability of 2495 hemagglutinins along the time course in Figure 6-6. Several features are highlighted from this figure:

(i) The amino-acid distribution probability has accurately and reliably recorded this evolutionary process from 1918 to 2006. This is the great advantage of our method [39, 42, 64].

(ii) The unequal means along the time course indicate the historical mutation process [39, 42, 64].

(iii) The fluctuating means allow us to use the fast Fourier transform to determine the periodicity, by which we can compare the patterns in stratified years as we have done in Chapter 4 [64, 82, 83].

(iv) The trend line suggests that the distribution rank of amino acids decreases slightly along the time course, which is identical to what we have seen in Figure 3-5 [64, 82, 83].

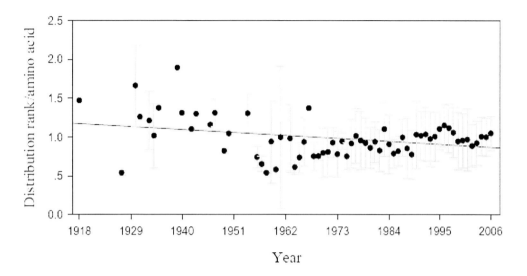

Figure 6-6. Distribution rank per amino acid of influenza A virus hemagglutinins from 1918 to 2006. The data are presented as mean±SD. The solid line is regressed line.

6.4. Meaning of Amino-Acid Distribution Probability

After observing the behavior of amino-acid distribution probability from spatial and time angles, we become more familiar with the amino-acid distribution probability in different scenarios, so that we can figure out its meanings.

The amino-acid distribution probability is a measure to gauge the randomness along a protein sequence, that is, the spatial measure. Thus, it can answer the question why a certain kind of amino acids does not homogenously distribute along a protein sequence, but clusters in different regions. In real world, it is very difficult to find out a kind of amino acids distributing homogenously in a protein. From the viewpoint of randomness, the homogenous distribution of amino acids along a protein sequence has a very small chance of occurrence, so the amino acids should distribute unequally. This suggests that the distribution probability governed by nature demands the amino acids to distribute heterogeneously along a protein sequence resulting in clusters and functional centers. This is the base for the high-level of protein structure and for protein functions.

From a probabilistic viewpoint, the larger the distribution probability is or the smaller the distribution rank is, the larger the randomness of amino-acid distribution is, the easier the construction of a protein is, and the more stable the protein is [63, 64, 83, 117-120].

Chapter 7

Application of Amino-Acid Distribution Probability

In this Chapter, we will mainly introduce how to apply the amino-acid distribution probability in research, thus this Chapter similar to Chapter 4 is mainly research-oriented.

We have seen in previous Chapters that the amino-acid distribution probability captures the randomness from the viewpoint that is different from the amino-acid pair predictability. However, both measures do the same things, say, they quantify a protein in whole as well as an individual amino acid in the protein. Therefore we can principally apply the same mathematical tools to deal with the quantified protein. For example, we have showed the application of fast Fourier transform to the historical curve of amino-acid pair predictability in Figure 3-5 to determine its periodicity [82]. We can also apply the fast Fourier transform to treating the historical curve of amino-acid distribution probability to determine its periodicity, which actually was done by us [83].

The advantage of our methods is that they can reliably record any slight change in a protein due to a mutation no matter of what cause initiates the mutation (for reviews see [39, 63, 64]). By quantifying historical proteins, we can get various patterns of evolutionary process of proteins [82, 83]. Reversibly, we can use these patterns to track the causes, which induced the change in evolutionary process of proteins [85].

Although we can apply all the procedures described in Chapter 4 to treating proteins and their mutations using the amino-acid distribution probability, we consider this treatment somewhat repeated because all the researchers can follow the procedures described in Chapter 4 to deal with similar problems using the amino-acid distribution probability. We would like to introduce the applications of the amino-acid distribution probability to study proteins and their mutations from different angles. Technically, we will use the amino-acid distribution rank more frequently because it is more suitable for these types of studies.

7.1. Mutation Effect on Amino-Acid Distribution

We would obtain a deep insight into the mutations if we could find the mutation effect on the original amino acids, mutated amino acids and a protein in whole.

Now we look at the mutation effect on human hemoglobin β-chain, of which 244 missense point mutations have been recorded. Historically, the human hemoglobin β-chain has been recorded in two forms: the record in several years ago contained 146 amino acids while the current one contains 147 amino acids. The difference between two forms is whether the hemoglobin begins from methionine "M". To stress the sensitivity of amino-acid distribution probability, we herein use the human hemoglobin β-chain with 146 amino acids for this Chapter.

Table 7-1 shows the amino-acid distribution probability in human hemoglobin β-chain with 146 amino acids. Compared Table 7-1 with Table 6-8, we can see that the only difference is related to methionine "M", including its number, predicted distribution probability and equal distribution probability in columns 2, 4 and 5, respectively. However, big differences can be found in the actual distribution probability and rank in five kinds of amino acids. These results once again demonstrate the sensitivity of amino-acid distribution probability.

As we have done in Chapter 4, we will study (i) the mutation effect on the amino acid targeted by mutations, (ii) the mutation effect on the amino acid appeared through mutations, and (iii) the overall effect of mutations on all amino acids in the protein (Figure 7-1).

Table 7-1. Actual and predicted distribution probability in human hemoglobin β-chain

Amino acid	Number	Actual distribution probability	Predicted distribution probability	Equal distribution probability	Rank	Rank/amino acid
A	15	0.0841	0.1569	2.9863e-6	3	0.2000
R	3	0.6667	0.6667	0.2222	1	0.3333
N	6	0.2315	0.3472	0.0154	2	0.3333
D	7	0.1071	0.3213	6.1199e-3	4	0.5714
C	2	0.5000	0.5000	0.5000	1	0.5000
E	8	0.2523	0.2523	2.4033e-3	1	0.1250
Q	3	0.6667	0.6667	0.2222	1	0.3333
G	13	0.1158	0.1544	2.0560e-5	2	0.1538
H	9	0.1967	0.1967	9.3666e-4	1	0.1111
I	0	- -	- -	- -	- -	- -
L	18	0.0831	0.1246	1.6272e-7	2	0.1111
K	11	7.6948e-3	0.2020	1.3991e-4	14	1.2727
M	1	1	1	1	1	1.0000
F	8	0.2243	0.2523	2.4033e-3	2	0.2500
P	7	0.3213	0.3213	6.1199e-3	1	0.1429
S	5	0.2880	0.3840	0.0384	2	0.4000
T	7	0.1285	0.3213	6.1199e-3	3	0.4286
W	2	0.5000	0.5000	0.5000	1	0.5000
Y	3	0.6667	0.6667	0.2222	1	0.3333
V	18	0.1246	0.1246	1.6272e-7	1	0.0556

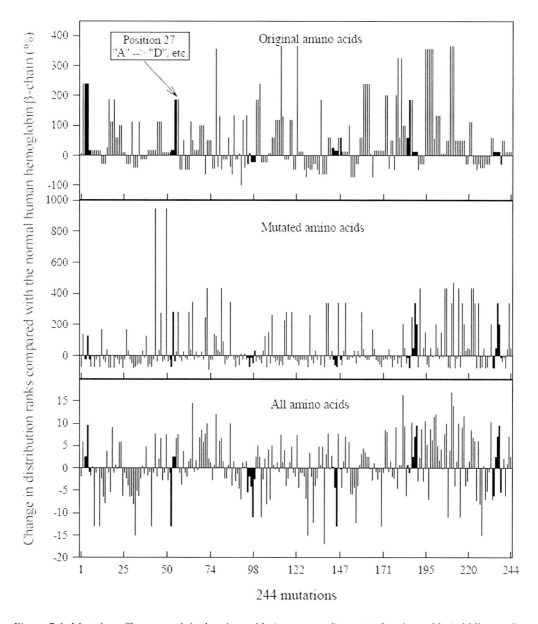

Figure 7-1. Mutation effects on original amino acids (upper panel), mutated amino acids (middle panel) and all amino acids in human hemoglobin β-chain (lower panel) from the viewpoint of amino-acid distribution rank.

7.1.1. Mutation Effect on Original Amino Acid

Figure 7-1 can be read as follows. The x-axis ranging from 0 to 245 represents 244 mutations in human hemoglobin β-chain, which are recorded from position 1 to position 146 and each position may have several different mutations. The horizon line crossing the zero in y-axis in each panel is referenced to the distribution rank in the normal human hemoglobin β-chain. Thus, each line standing on the zero-line in each panel represents a mutation, so there are 244 lines representing 244 mutations in each panel. The direction of line indicates whether a mutation leads to the increase or decrease in the amino-acid distribution rank of original amino acid (the upper panel), mutated amino acid (the middle panel) and all amino acids (the lower panel) in the mutant hemoglobin in comparison with that of normal human hemoglobin β-chain (zero-line). The height of line indicates the magnitude of change in the distribution rank induced by a mutation.

For example, three mutations have been documented at position 27 changing alanine "A" to aspartic acid "D", serine "S" and valine "V", respectively. Table 7-2 shows the distribution configuration of alanines "A" before and after mutation in human hemoglobin β-chain, for which we once again can calculate the distribution probability using Equation 1 in Chapter 5.

Table 7-2. Distribution of alanines "A" in human hemoglobin β-chain before and after mutation

Part	I	II	III	IV	V	VI	VII	VIII	IX	X	XI	XII	XIII	XIV	XV
Before mutation	1	1	1	0	0	1	2	1	1	0	0	1	2	3	1
After mutation	1	1	0	0	1	1	2	1	0	0	1	2	4	0	- -

The distribution probability before mutation is

$$\frac{15!}{1!\times1!\times1!\times0!\times0!\times1!\times2!\times1!\times1!\times0!\times0!\times1!\times2!\times3!\times1!}\times$$

$$\frac{15!}{4!\times8!\times2!\times1!\times0!\times0!\times0!\times0!\times0!\times0!\times0!\times0!\times0!\times0!\times0!}\times15^{-15}=0.0841$$

So the distribution probability for 15 alanines "A" is 0.0841, which is ranked as 3 because there are tow distribution probabilities ahead (0.1569 and 0.0981). Consequently, the distribution rank per alanine "A" is 0.2 (3/15 = 0.2).

After mutation, the distribution probability is

$$\frac{14!}{1!\times1!\times0!\times0!\times1!\times1!\times2!\times1!\times0!\times0!\times1!\times2!\times4!\times0!}\times$$

$$\frac{14!}{5!\times6!\times2!\times0!\times1!\times0!\times0!\times0!\times0!\times0!\times0!\times0!\times0!\times0!}\times14^{-14}=0.0412$$

Thus the distribution probability for the left 14 alanines "A" is 0.0412 and ranked as 8, and the distribution rank per alanine is 0.5714 (8/14 = 0.5714).

The mutation at position 27 leads the distribution rank of alanines changes to 0.5714 from 0.2, which equal to the increase of 185.71% ((0.5714 – 0.2)/0.2). Thus, the line height for the mutation at position 27 is 185.71% in comparison with the normal human hemoglobin β-chain in the upper panel of Figure 7-1. As three mutations at position 27 target the same

"A", so the changes are the same in their distribution ranks, and there are three lines with the same height in the upper panel.

Hence, each line in the upper panel of Figure 7-1 tells us the effect of each mutation on the distribution rank, that is, a mutation can either increase or decrease the distribution rank of the kind of original amino acids. It is likely that most mutations in human hemoglobin β-chain lead to the increase in the distribution rank of original amino acids.

7.1.2. Mutation Effect on Mutated Amino Acid

The middle panel of Figure 7-1 illustrates the changes in distribution rank of mutated amino acids. Again, we can see the example at position 27, where alanine "A" is changed to aspartic acid "D", serine "S" and valine "V", respectively.

Before mutation, seven aspartic acids "D" distribute in the pattern displayed in Table 7-3, but their distribution pattern is changed after mutation.

Table 7-3. Distribution of aspartic acids "D" in human hemoglobin β-chain before and after mutation

Part	I	II	III	IV	V	VI	VII	VIII
Before mutation	1	0	2	2	2	0	0	- -
After mutation	0	2	2	1	2	1	0	0

The distribution probabilities before and after mutation are

$$\frac{7!}{1!\times0!\times2!\times2!\times2!\times0!\times0!} \times \frac{7!}{3!\times1!\times3!\times0!\times0!\times0!\times0!\times0!} \times 7^{-7} = 0.1071$$

and

$$\frac{8!}{0!\times2!\times2!\times1!\times2!\times1!\times0!\times0!} \times \frac{8!}{3!\times2!\times3!\times0!\times0!\times0!\times0!\times0!} \times 8^{-8} = 0.1862,$$

and their distribution ranks are 4 and 3, respectively. Thus, the distribution rank per aspartic acid "D" is 0.5714 (4/7 = 0.5714) before mutation and 0.375 (3/8 = 0.375) after mutation, which leads to −34.38% ((0.375 − 0.5714)/0.5714) change in the distribution rank.

These changes can be found at the corresponding place in the middle panel, and each line indicates the mutation effect on the distribution rank of mutated amino acids.

7.1.3. Mutation Effect on Whole Protein

Finally, we see the mutation effect on human hemoglobin β-chain in whole (the lower panel of Figure 7-1). We still look at the corresponding place in the lower panel, namely, at position 27, where we have recorded the mutation from alanine "A" to aspartic acid "D", and we already know the changes in the distribution ranks for alanine "A" and aspartic acid "D" before and after mutation.

Table 7-4 lists the distribution ranks for all 20 kinds of amino acids before and after the mutation changing alanine "A" to aspartic acid "D" at position 27. At the bottom row, we can see that the totals for rank/amino acid are 7.1556 and 7.3305 before and after mutation, whose difference is the effect of this mutation on the whole human hemoglobin β-chain. We present this difference as a line of 2.44% height ((7.3305 − 7.1556)/7.1556) in the lower panel of Figure 7-1. In such a manner, we can see the effect of each mutation on the distribution rank of human hemoglobin β-chain in whole.

Table 7-4. Distribution rank in human hemoglobin β-chain before and after mutation changing alanine "A" to aspartic acid "D" at position 27

Amino acid	Before mutation			After mutation		
	Number	Rank	Rank/amino acid	Number	Rank	Rank/amino acid
A	15	3	0.2000	14	8	0.5714
R	3	1	0.3333	3	1	0.3333
N	6	2	0.3333	6	2	0.3333
D	7	4	0.5714	8	3	0.3750
C	2	1	0.5000	2	1	0.5000
E	8	1	0.1250	8	1	0.1250
Q	3	1	0.3333	3	1	0.3333
G	13	2	0.1538	13	2	0.1538
H	9	1	0.1111	9	1	0.1111
I	0	- -	- -	0	- -	- -
L	18	2	0.1111	18	2	0.1111
K	11	14	1.2727	11	14	1.2727
M	1	1	1.0000	1	1	1.0000
F	8	2	0.2500	8	2	0.2500
P	7	1	0.1429	7	1	0.1429
S	5	2	0.4000	5	2	0.4000
T	7	3	0.4286	7	3	0.4286
W	2	1	0.5000	2	1	0.5000
Y	3	1	0.3333	3	1	0.3333
V	18	1	0.0556	18	1	0.0556
Total	146	44	7.1556	146	48	7.3305

Table 7-5 summarizes up the effects of all 244 mutations in three panels of Figure 7-1. It can be seen that about a half of mutations lead to the decrease in the distribution ranks in human hemoglobin β-chain in whole (the lower panel). As we mentioned in Chapter 6, the smaller the rank is, the more random a protein is constructed. Hence, we can say that about a half of 244 mutations in human hemoglobin β-chain can be explained by random mechanism or are initiated by randomness. Now we can draw some implications from Figure 7-1 and Table 7-5 as follows.

(i)   Most mutations lead to the increase in amino-acid distribution rank of original amino acids, as can be seen that most lines are above the zero line in the upper panel. The increase in distribution rank means that the structure of mutant human hemoglobin β-chain becomes less random and more complicated, which could be the base for function. Thus, the upper panel suggests that nature eliminates unnecessary amino acid through mutation.

(ii) However, most mutated amino acids in fact decrease the distribution rank because most lines are below the zero line in the middle panel. The decrease in distribution rank means that the structure of mutant human hemoglobin β-chain becomes more random and less complicated, which could not be necessary for function. Thus, the middle panel suggests that nature somewhat randomly picks up an amino acid as a mutated amino acid according to parsimony.

(iii) Similar numbers of increase and decrease in the distribution rank are found in the lower panel, which suggests that nature engineers mutations but cannot predict what overall effect for each mutation, i.e. nature cannot predict the future. Thus, the protein evolution can continue because mutations can either increase or decrease the randomness in a protein [63, 64, 83, 117-120].

Furthermore, we can get more insight if we analyze a particular mutation in Figure 7-1 in much greater details.

Table 7-5. Effect of 244 mutations on amino-acid distribution rank of human hemoglobin β-chain presented in Figure 7-1

Distribution rank/Amino acid	Upper panel	Middle panel	Lower panel
Increase	162	90	126
Unchanged	0	5	0
Decrease	82	149	118

7.2. Distribution Rank and Protein Structure/Function

One of hot issues in protein science is the structure-function relationship [121, 122]. Meanwhile, a mutation changes the structure of a protein, this may result in a changed function. One might wonder if we can establish the relationship between amino-acid distribution rank and protein function. However, we are not quite sure on this issue because the relationship between a particular protein function and a measure is generally based on the measure that is specifically designed and developed for the function. For example, pH is specifically designed and developed for measuring the degree of acid and base. On the other hand, our measures are developed for measuring randomness in a protein, which could have some indirect relationship with some particular functions.

7.2.1. Distribution Rank and Structural Stability

Among 244 missense point mutations in human hemoglobin β-chain, 159 (159/244 = 65.16%) mutations do not change the stability of mutant hemoglobin, but the rest 85 (85/244 = 34.84%) mutations lead to the mutant hemoglobin unstable. This means that each mutation has about 1/3 chance of resulting in the mutant human hemoglobin β-chain unstable.

We can analyze this issue by classifying the data in the lower panel of Figure 7-1 according to whether the mutation leads to the mutant hemoglobin stable or unstable. Then we present these classified data in Figure 7-2. The most pronounced difference between the upper and lower panels is that the distribution of stable mutants is quite symmetric, where 85 and 74 mutants are below and above the zero horizon line, but the distribution of unstable mutants is asymmetric, where 33 and 52 mutants are below and above the zero horizon line. Overall, there are about a half of mutations below the zero horizon line (85 + 33 = 118), and about a half above the zero horizon line (74 + 52 = 126).

The results are very suggestive, that is, a mutation would have a larger chance of leading the mutant hemoglobin β-chain unstable if the mutation brings about an increase in amino-acid distribution rank. This is in good agreement with our notation that the larger the distribution rank is, the more unstable the protein is, or in words of distribution probability, the smaller the distribution probability is, the more unstable the protein is [63, 64, 83, 117-120]. This way, the quantification of randomness developed by us does have a relationship with the structural stability of proteins.

In addition, the mutations, which do not affect the stability of mutant hemoglobin, produce a more random distribution in the upper panel in the form of symmetry.

7.2.2. Distribution Rank and Function

An important measure for hemoglobin function is the affinity to oxygen. Among 244 missense point mutations in human hemoglobin β-chain, 61 (61/244 = 25%), 159 (159/244 = 65.16%) and 24 (24/244 = 9.84%) mutations lead the affinity up, unchanged and down, respectively, in mutant human hemoglobin β-chains. Therefore, the oxygen affinity of mutant hemoglobin would have 2/3 chance of being unchanged, 1/4 up and 1/10 down.

As seen in Figure 7-3, the distribution in the upper panel is asymmetric, where 23 and 38 mutants are below and above the zero horizon line, the distribution in the middle panel is symmetric, where 80 and 79 mutants are below and above the zero horizon line, and the distribution in the lower panel is asymmetric, where 15 and 9 mutants are below and above the zero horizon line.

Without detailed elaborations, we can know that an increased distribution rank is more related to increase in oxygen affinity, but a decreased distribution rank is more related to decrease in oxygen affinity.

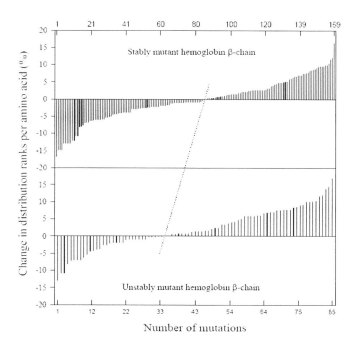

Figure 7-2. Distributions of change in amino-acid distribution rank with respect to the mutant human hemoglobin β-chain stable and unstable.

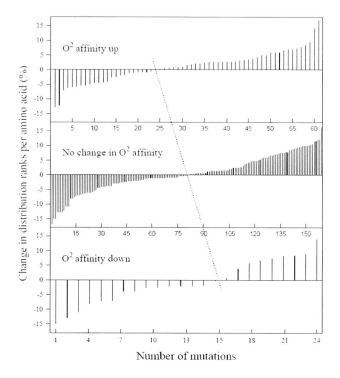

Figure 7-3. Distributions of change in amino-acid distribution rank with respect to the oxygen affinity being up, unchanged and down in mutant human hemoglobin β-chain.

Once again, the relationship between the oxygen affinity and distribution rank of amino acids confirms our logic, that is, the bigger the distribution rank is, the more complicated the protein structure is, the better function the protein has. Opposite is held for the lower panel of Figure 7-3.

7.3. Cross-Impact Analysis

Until now, we have used our methods to deal with mutation problems from computational angle. What can we still achieve using our quantitative methods? There are two questions requiring a considerable effort in mutation filed: One is the mutation probability for a protein in question, and the other is the extent to which a mutation cause promotes mutation. In this section, we will apply another mathematical tool, cross-impact analysis, to addressing these two questions with help of the amino-acid distribution probability [123, 124].

7.2.1. Cross-Impact Analysis, Bayes' Law and Mutation Probability

Although we generally believe that a mutation would have a cause, intuitively and empirically, we know that a cause can either lead to a mutation or not, whereas a mutation can occur without any particular cause. Still, a mutant protein can either lead to a disease or not, while a disease can exist with the normal protein. These yes-and-no coupling relationships can be presented using the cross-impact analysis in Figure 7-4 [125-129].

At the impact level, $P(2)$ is the probability that the mutation cause appears, and $P(\bar{2})$ is the probability that the mutation cause does not appear.

At the mutation level, $P(\bar{1}\,|\,2)$ is the impacted probability that the mutation does not occur when its cause appears, $P(1\,|\,2)$ is the impacted probability that the mutation occurs when its cause appears. $P(\bar{1}\,|\,\bar{2})$ is the impacted probability that neither the mutation nor its cause occurs, and $P(1\,|\,\bar{2})$ is the impacted probability that the mutation occurs without the defined cause.

At the influenza level, $P(\bar{1}\,2)$ or $P(\bar{1}\,\bar{2})$ is the probability that influenza does not occur when the mutation cause appears or does not appear, whereas $P(1\,2)$ or $P(1\,\bar{2})$ is the probability that influenza occurs when the mutation cause appears or does not appear.

As seen in Figure 7-4, the cross-impact analysis is a probabilistic approach, which, although defined in such a name, in fact uses the conditional probability for analysis and calculation. We can trace the process from impact to mutation to influenza through Figure 7-4. For example, the probability $P(1\,2)$ will increase if an impact enhances the chance of occurring of mutation.

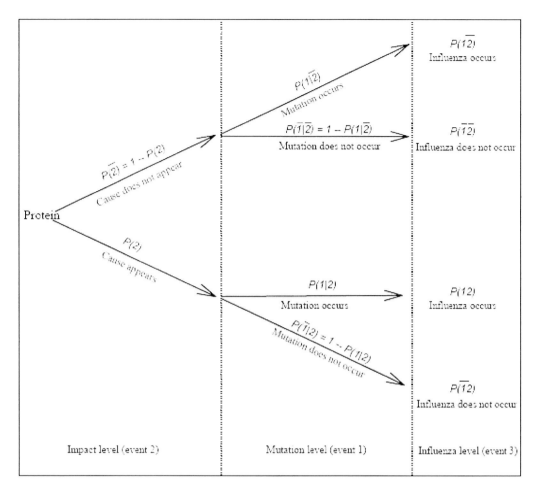

Figure 7-4. Cross-impact relationship among mutation, mutation cause and influenza.

Because we have just studied the problem of the change in distribution rank versus the stability of mutant human hemoglobin β-chain with complete dataset in Section 7.2.1, we can similarly format these data into the cross-impact analysis. Figure 7-5 not only gives us much more sense on the numerical relationship of cross-impact analysis, but also makes us be familiar with its calculation (Table 7-6). Moreover, we can see whether the results from cross-impact analysis can provide an identical conclusion as that in Section 7.2.1.

Several features can be drawn from Figure 7-5 and Table 7-6:

1. As $P(\overline{2})$ is larger than $P(2)$, a mutation has a larger chance of increasing the distribution rank in mutant hemoglobin β-chain.

2. As $P(1\,|\,\overline{2})$ is larger than $P(\overline{1}\,|\,\overline{2})$, a mutation increasing the distribution rank has a larger chance of leading to the mutant hemoglobin stable.

3. As $P(1\,|\,2)$ is much larger than $P(\overline{1}\,|\,2)$, a mutation decreasing the distribution rank has a far larger chance of resulting in a stable mutant hemoglobin.

4. As $P(1|\overline{2})$ and $P(1|2)$ are larger than $P(\overline{1}|\overline{2})$ and $P(\overline{1}|2)$, a mutation changing the distribution rank is more likely to bring about a stable mutant hemoglobin.

Table 7-6. Calculation on cross-impact analysis in Figure 7-5

$P(2) = 118/244 = 0.4836$
$P(\overline{2}) = 1 - P(2) = 0.5164 = 126/244$
$P(1|\overline{2}) = 74/126 = 0.5873$
$P(\overline{1}|\overline{2}) = 52/126 = 0.4127$
$P(1|2) = 85/118 = 0.7203$
$P(\overline{1}|2) = 33/118 = 0.2797$
$P(1\overline{2}) = P(1|\overline{2}) \times P(\overline{2}) = 74/126 \times 0.5164 = 0.3033 = 74/244$
$P(\overline{1}\,\overline{2}) = P(\overline{1}|\overline{2}) \times P(\overline{2}) = 52/126 \times 0.5164 = 0.2131 = 52/244$
$P(12) = P(1|2) \times P(2) = 85/118 \times 0.4836 = 0.3484 = 85/244$
$P(\overline{1}2) = P(\overline{1}|2) \times P(2) = 33/118 \times 0.4836 = 0.1352 = 33/244$

Very closely related to the application of cross-impact analysis is the Bayes' law [130], which indicates that the probabilities of occurrence of two events can be related by

$$P(1|2) = \frac{P(2|1)}{P(2)} P(1) \hspace{4cm} \text{Equation 2}$$

In Equation 2, we can see that $P(2)$ and $P(1|2)$ have been defined in cross-impact analysis, but $P(1)$ and $P(2|1)$ have not. Now let us see how we can define them.

From Figure 7-5, we know that $P(2)$ is the probability that the rank decreases in mutant human hemoglobin β-chain, and $P(1|2)$ is the probability that the mutant hemoglobin is stable with decreased rank. Thus, $P(1)$ is the probability that the mutant human hemoglobin β-chain is stable, and $P(2|1)$ is the probability that the rank decreases under the condition that the mutant hemoglobin is stable.

Again from Figure 7-5, we have $P(1|2) = 85/118 = 0.7203$ and $P(2|1) = 85/(85 + 74) = 0.5346$, so we have $P(1) = \dfrac{P(1|2)}{P(2|1)} P(2) = \dfrac{0.7203 \times 0.4836}{0.5346} = 0.6516$. This $P(1)$ indicates that the hemoglobin is relatively stable to a variety of various turbulence because $P(1) > 0.5$, thereafter we would expect to see most mutant hemoglobins stable.

However, the meaning of $P(1)$ is much different for Figure 7-4, because the cross-impact analysis defines different events from those in Figure 7-5.

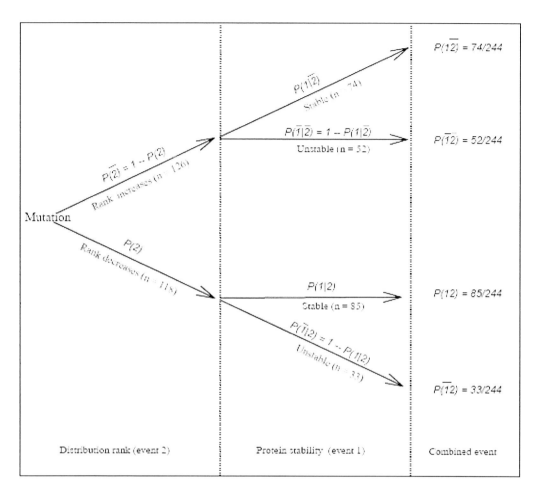

Figure 7-5. Cross-impact relationships among distribution rank, stability of mutant human hemoglobin β-chain and combined events based on the data in Section 7.2.1.

In the sense of Figure 7-4, $P(1)$ is the probability of spontaneously occurring of mutation, which is remarkable as we have no other efficient way to find the spontaneous mutation probability $P(1)$ in various contexts. After detailed elaboration on Figure 7-5, we can consider how to apply the cross-impact analysis and Bayes' law to the coupling relationship in Figure 7-4. In mathematical modeling, the first issue is if we have the data suited for the given mathematical model. For the proteins from influenza A virus, we have 2495 computed hemagglutinins (Table 3-5), of which H5 is the confirmed subtype with the outbreak of bird flu and influenza in 1997 [58-62], thus, H5 hemagglutinins are our focus for cross-impact analysis.

Apparently, our data still cannot meet the demand by cross-impact analysis, because we do not know how many causes there are, how many mutations are attribute to certain causes, etc. We therefore need to find out ways to overcome theses difficulties.

1. We know that there was an outbreak of bird flu in 1997, which can mainly be considered as the cause-driven-event. This means that the amino-acid distribution

rank (amino-acid pair predictability as well) in some hemagglutinins would be different from the mean value obtained from all documented hemagglutinins in 1997, which is $P(1|2)$, the probability that the mutation occurs when its cause appears in Figure 7-4.

2. We can find that the distribution rank of some hemagglutinins is different from the baseline level obtained from all documented hemagglutinins over a period of time. These can be considered as the occurrence of mutations with and without their causes, which is $P(1|\overline{2})+P(1|2)$ in Figure 7-4.

3. From $P(1|\overline{2})+P(1|2)$, we pick out the hemagglutinins in 1997, which is $P(2|1)$, the probability that the occurrence of mutations can be related to their causes in the Bayes's law (Equation 2).

4. Again, we specify 1997 as the year when the mutation cause appeared, which is certainly not an event appeared each year. In Section 4.2, we have used the fast Fourier transform to stratify the amino-acid pair predictability along the time course to find the periodicity [83, 83]. Theoretically, the cause in 1997 would belong to a periodicity, such as 7.3 years. Then we have $P(2)$, the probability that the mutation cause appears.

5. In such a manner, we can compute the spontaneous mutation probability $P(1)$ using Bayes' Law (Equation 2).

Accordingly, we can conduct the computation based on the above five points, Table 7-7 details the calculation process. This table can be divided into four parts. The upper-left part is the statistical data for the hemagglutinins of H5N1 influenza viruses from 1996 to 2002, which is a periodicity determined using the fast Fourier transform. The upper-right part is the counted numbers of hemagglutinins, which are either above or below the mean of the distribution rank in the given year/years. The lower-left part is the calculations according to the cross-impact analysis and Bayes' law. The lower-right part is the explanations for the calculations in the corresponding rows of the lower-left part.

As can be seen, $P(1)$ is 0.21, i.e. the mutation probability for H5N1 hemagglutinins is about 0.2 per generation. This probability can be viewed as the probability of spontaneous occurrence of mutation, while a mutation cause will increase this probability [64, 124].

Table 7-7. H5N1 hemagglutinins used in calculations for cross-impact and Bayes' law

Year	Distribution rank		Number of H5N1 hemagglutinins		
	Mean	SD	Total	Below mean	Above mean
1996	1.2610	0.0859	2	1	1
1997	1.1320	0.1340	34	12	22
1998	1.0420	0.2040	2	1	1
1999	1.3190	0.0640	5	2	3
2000	1.1570	0.2790	23	11	12
2001	1.1960	0.2050	37	13	24
2002	1.1120	0.2060	39	18	21

Year	Distribution rank		Number of H5N1 hemagglutinins		
	Mean	SD	Total	Below mean	Above mean
1996—2002	1.1545	0.2031	142	58	84
Excluding 1997	1.1616	0.2206	108	60	48

$P(1\,\vert\,2) = 22/34 = 0.65$	Refers to hemagglutinin number in 1997
$P(1\,\vert\,\overline{2}) + P(1\,\vert\,2) = 84/142 = 0.59$	Refers to hemagglutinin number in 1996—2002
$P(2\,\vert\,1) = 48/108 = 0.44$	Refers to hemagglutinin number excluding 1997
$P(2) = 1/7 = 0.14$	Refers to outbreak of flu in this 7-year periodicity
$P(1) = 0.65/0.44 \times 0.14 = 0.21$	Refers to the Bayes' Law (Equation 2)

7.2.2. Enhancement and Inhibition

In the last section, we mentioned that the mutation cause could increase the mutation probability, which seems to be the common sense.

The cross-impact analysis has defined both enhancement and inhibition with respect to coupled events [129]. The enhancement is that the occurrence of the second event enhances the probability of occurrence of the first event, say, $P(1\,\vert\,2) > P(1)$, for our study, it is a cause that increases the occurrence of mutation. The inhibition is that the occurrence of second event inhibits the probability of the occurrence of the first event, say, $P(1\,\vert\,2) < P(1)$, for our study, that is a cause that decreases the occurrence of mutation. As we already said that we generally consider that the mutation cause increases the chance of occurrence of mutation, we have the following equations to estimate the extent, to which the mutation cause promotes the mutation.

$$1 - \frac{1 - P(1)}{1 - P(2)} \leq P(1\,\vert\,\overline{2}) \leq P(1) \qquad\qquad\qquad \text{Equation 3}$$

$$P(1) \leq P(1\,\vert\,2) \leq \frac{P(1)}{P(2)} \qquad\qquad\qquad \text{Equation 4}$$

For Equations 3 and 4, we already know the meanings and values of all the probabilities from cross-impact analysis and Bayes' law. The middle term in Equation 4 is the impacted probability under the influence of mutation cause, while the middle term in Equation 3 presents the opposite case that the impacted probability without influence of mutation cause. Still, the left and middle terms in Equation 4 shows the cross-impact enhancement, $P(1\,\vert\,2) > P(1)$. This also indicates that $P(1)$ is the spontaneous mutation probability without any impact. On the other hand, Equations 3 and 4 actually give the range or bound of enhancement, for example, $P(1\,\vert\,\overline{2})$ cannot be larger than $P(1)$.

After understanding the meanings of Equations 3 and 4, we look at what we have for calculation of enhancement and how to visualize the enhancement. First, we have $P(1)$ by calculating the Bayesian equation. For a well-recorded dataset, we have $P(1|2)$ and $P(1|\overline{2})$, which although is a very rare case. In such a case, we can use different values of $P(2)$ to observe its impact on $P(1|2)$ and $P(1|\overline{2})$.

Figure 7-6 illustrates the impacted probabilities with respect to the probability of occurring of mutation $P(1)$ and the probability of appearing of cause $P(2)$ in the hemagglutinins of H5 influenza viruses. This figure can be read as follows. First, the left panel shows these three probabilities in a 3-dementional configuration: $P(1)$ is in x-axis, $P(2)$ is in z-axis, and $P(1|\overline{2})$ and $P(1|2)$ are in y-axis, because three probabilities included in Equations 3 and 4. Then the calculatedly impacted probabilities $P(1|\overline{2})$ and $P(1|2)$ are presented by triangles ABC and ACD, respectively.

Although the current graphic software is powerful in making 3-dimensional graph, the 2-dimensional graph can give us more precise evaluation after certain transform [64, 124]. Thus, we transfer this 3-dimensional figure into a 2-dementional figure (the right panel), so the dynamics of impact on the occurring of mutation can be viewed simply and clearly. In the right panel of Figure 7-6, as the chance of appearing of mutation cause increases or the intensity of $P(2)$ increases along y-axis, the probability of occurring of spontaneous mutations decreases [the black triangle $P(1|\overline{2})$] whereas the probability of occurring of induced mutations increases [the gray triangle $P(1|2)$]. Also, the impact can significantly enhance the probability of occurring of mutations $P(1)$, whose range enlarges to the up-right corner from the down-left one. Numerically, the $P(1)$ of H5 hemagglutinins changes to the range of $0.79 - 1$ from that of $0 - 0.21$.

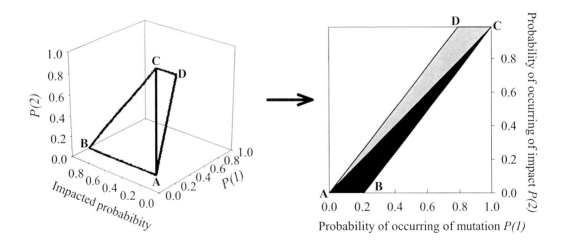

Figure 7-6. Enhancement of mutation cause on the occurrence of mutations.

Chapter 8

Future Amino-Acid Composition

This Chapter is somewhat different from the previous ones, because we will not develop a new measure to gauge the randomness in a protein. This is so because our previous two measures, i.e. amino-acid pair predictability and amino-acid distribution probability, deal with the spatial randomness in general, which are sufficient to gauge the randomness in the primary structure, a linear structure, of a protein.

However, the implications in previous chapters are that the spatial randomness can engineer mutations, which occur in the next generation of proteins. This further means that the spatial randomness is time-oriented, forwards the next generation. This inspires us if there is a random measure that is time-oriented.

At amino-acid level, the issue with time-oriented is that an amino acid mutates to another amino acid, which seems to be a somewhat old problem, because several approaches have been developed [131-134] to analyze the problem of an amino acid mutating to another one, even with probability. Among them, the PAM matrix is most prominent, which is developed based on a database of 1,572 changes in 71 groups of closely related proteins [131], thus this PAM matrix can be viewed as empiric.

One might wonder why we are interested in this problem, which seems to be solved through several ways. The main reason is that these methods are mainly based on the experimental data, i.e. empiric model, and are exclusively related to amino acids rather than RNA codons, whereas we are more theoretical-oriented.

8.1. Translation Probability between RNA Codons and Translated Amino Acids

We know that the amino acids have the unambiguous relationship with RNA codons (Table 8-1), which, however, can tell more than we previously noticed.

Table 8-1. 64 RNA codons and their translated amino acids

UUU UUC	Phenylalanine, F	UCU UCC	Serine, S	UAU UAC	Tyrosine, Y	UGU UGC	Cysteine, C
UUA UUG	Leucine, L	UCA UCG		UAA UAG	STOP	UGA UGG	STOP Tryptophan, W
CUU CUC CUA CUG	Leucine, L	CCU CCC CCA CCG	Proline, P	CAU CAC	Histidine, H	CGU CGC CGA CGG	Arginine, R
				CAA CAG	Glutamine, Q		
AUU AUC AUA	Isoleucine, I	ACU ACC ACA ACG	Threonine, T	AAU AAC	Asparagine, N	AGU AGC	Serine, S
AUG	Methionine, M			AAA AAG	Lysine, K	AGA AGG	Arginine, R,
GUU GUC GUA GUG	Valine, V	GCU GCC GCA GCG	Alanine, A	GAU GAC	Aspartic acid, D	GGU GGC GGA GGG	Glycine, G
				GAA GAG	Glutamic acid, E		

In Table 8-1, an amino acid can be translated by different RNA codons because the number of RNA codons is larger than that of amino acids. For example, methionine "M" is related to a single RNA codon AUG, phenylalanine "F" is related to two RNA codons UUU and UUC, isoleucine "I" is related to three RNA codons AUU, AUC and AUA, proline "P" is related to four RNA codons CCU, CCC, CCA and CCG, and leucine "L" is related to six RNA codons UUA, UUG, CUU, CUC, CUA and CUG.

What does this observation mean? We know that there are four RNA codes, A, C, G, and U. If one of four codes mutates, what will happen at amino-acid level for the order, methionine, phenylalanine, isoleucine, proline and leucine? We can deduce the order in such a manner: as the mutation of a single RNA code will change a single RNA codon, then methionine will certainly mutate to another amino acid because methionine is only related to a single RNA codon, but phenylalanine has a smaller chance of mutating because it is related to two RNA codons, followed by isoleucine because it is related to three RNA codons, then proline because it is related to four RNA codons, and finally leucine because it is related to six RNA codons. This is so because we do not know whether the mutated RNA code is exactly the RNA codon that encodes the phenylalanine, isoleucine, proline and leucine. This example implies the time-oriented randomness.

Now we begin to deduce the probability of time-oriented randomness. First, we focus on point mutation at a single RNA code, that is, the mutation only changes a RNA code, but does not insert or delete a RNA code, so the length of RNA sequence is not changed by mutation, except for the nonsense point mutation. Second, we list what happens when a point mutation occurs at the first, second and third position of a RNA codon.

For example, RNA codon AUG corresponds to amino acid methionine. If a mutation occurs at the first position A of the codon, A can mutate to A, U, G and C although A changes to A cannot be seen unless using isotope A. So AUG will mutate to AUG, UUG, GUG, and CUG, which correspond to methionine, leucine, valine and leucine according to Table 8-1. Similarly, we can list all changes in the first, second and third positions of the codon one by one, and the results are listed in Table 8-2. After counting the numbers of different translated amino acids in the last column in Table 8-2, we can find three methionines among 12 translated amino acids. These results construct the translation probability because this probability reflects the chance for amino acids to be translated from their RNA codons.

On the other hand, we may consider the self-oriented mutation at RNA code unnecessary, that is, we will not consider the cases where AUG mutates to AUG (all italic RNA codons in Table 8-2). Therefore, we can omit these three methionines in counting, and have a total of nine probabilities (the last row in Table 8-2).

Table 8-2. Methionine and its mutated amino acids

RNA codon				Translated amino acid
First position	Second position	Third position		
Change in the first position				
A	*U*	*G*	→	*Methionine*
U	U	G	→	Leucine
G	U	G	→	Valine
C	U	G	→	Leucine
	Change in the second position			
A	*U*	*G*	→	*Methionine*
A	A	G	→	Lysine
A	G	G	→	Arginine
A	C	G	→	Threonine
		Change in the third position		
A	*U*	*G*	→	*Methionine*
A	U	A	→	Isoleucine
A	U	U	→	Isoleucine
A	U	C	→	Isoleucine
Translated amino acids: 3 methionines + 2 leucines + 1 valine + 1 lysine + 1 arginine + 1 threonine + 3 isoleucines Translation probability: 3/12 + 2/12 + 1/12 + 1/12 + 1/12 + 1/12 + 3/12				
Translated amino acids without self-oriented RNA code mutation: 2 leucines + 1 valine + 1 lysine + 1 arginine + 1 threonine + 3 isoleucines Translation probability: 2/9 + 1/9 + 1/9 + 1/9 + 1/9 + 3/9				

Italics indicate the self-oriented mutation in RNA codon and its translated amino acid.

Table 8-2 means that a point mutation of RNA codon AUG has 3/12 chance of mutating to methionine, 2/12 chance of mutating to leucine, 1/12 chance of mutating to valine, 1/12 chance of mutating to lysine, 1/12 chance of mutating to arginine, 1/12 chance of mutating to threonine and 3/12 chance of mutating to isoleucine. If we do not consider the case of self-oriented mutation in RNA codon AUG to RNA codon AUG, we know that a point mutation of RNA codon AUG has 2/9 chance of mutating to leucine, 1/9 to valine, 1/9 to lysine, 1/9 to arginine, 1/9 to threonine and 3/9 to isoleucine.

In the same manner, we can determine all the translation probabilities from RNA codons to their translated amino acids, and in fact we determine these translation probabilities excluding self-oriented mutations in RNA codons [136].

8.2. Amino-Acid Mutating Probability

On the other hand, Table 8-2 only tells us the situation when a point mutation occurs in RNA codon we can estimate the probability that which type of amino acids would be likely to mutate. This seems too indirect because the development of our approaches currently is going at amino-acid level, and we are more interested in the mutation probability from an amino acid to other amino acids.

With the thought of translation probability from RNA codons to translated amino acids, we can deduce the mutation probability from an amino acid to another. However, this time we begin with an amino acid and its RNA codon rather than only RNA codon. In Table 8-3, we detail the possibility of threonine mutating to other amino acids based on the single point mutation at different position of RNA codons ACU, ACC, ACA and ACG. As the probability in the bottom row is relevant to the chance for an amino acid mutating to another one, we call it as the amino-acid mutating probability [135-137].

Naturally, Table 8-3 is the detailed deduction process for amino acid threonine. We can use the same manner to deduce all 20 kinds of amino acids. In addition, we are more interested in the probability, which excludes the self-oriented mutation at RNA code because such a mutation can only be detected using isotope. Table 8-4 shows the amino-acid mutating probability for all 20 kinds of amino acids, however please note that the letters in Table 8-4 represent amino acids rather than RNA codes.

8.3. Future Amino-Acid Composition

After having the amino-acid mutating probability, we can return to our initial aim of this chapter, that is, to develop a measure for time-oriented randomness. From Tables 8-3 and 8-4, we know that an amino acid has different probabilities of mutating to other amino acids, which are governed by a point mutation at its RNA codon. Now let us look at what we can get from the amino-acid mutating probability in Table 8-4.

Table 8-3. Amino acid threonine and its mutated amino acids according to a single mutation at different positions in four RNA codons ACU, ACC, ACA and ACG

Mutation position	RNA codon and its translated amino acid							
	ACU	Threonine	ACC	Threonine	ACA	Threonine	ACG	Threonine
The first position	UCU	Serine	UCC	Serine	UCA	Serine	UCG	Serine
	CCU	Proline	CCC	Proline	CCA	Proline	CCG	Proline
	GCU	Alanine	GCC	Alanine	GCA	Alanine	GCG	Alanine
The second position	AUU	Isoleucine	AUC	Isoleucine	AUA	Isoleucine	AUG	Methionine
	AAU	Asparagine	AAC	Asparagine	AAA	Lysine	AAG	Lysine
	AGU	Serine	AGC	Serine	AGA	Arginine	AGG	Arginine
The third position	ACA	Threonine	ACA	Threonine	ACU	Threonine	ACU	Threonine
	ACC	Threonine	ACU	Threonine	ACC	Threonine	ACA	Threonine
	ACG	Threonine	ACG	Threonine	ACG	Threonine	ACC	Threonine
Mutation results	4 alanines + 2 arginines + 2 asparagines + 3 isoleucines + 2 lysines + methionine + 4 prolines + 6 serines+ 12 threonines							
Mutation probability	4/36 + 2/36 + 2/36 + 3/36 + 1/36 + 4/36 + 6/36 + 12/36							

Table 8-4. Amino-acid mutating probability including self-oriented mutation at amino-acid level

Original amino acid	Future composition of amino acids with their mutating probability
A	12/36A+2/36D+2/36E+4/36G+4/36P+4/36S+4/36T+4/36V
R	18/54R+2/54C+2/54Q+6/54G+2/54H+1/54I+4/54L+2/54K+1/54M+4/54P+6/54S+2/54T+2/54W+2/54STOP
N	2/18N+2/18D+2/18H+2/18I+4/18K+2/18S+2/18T+2/18Y
D	2/18A+2/18N+2/18D+4/18E+2/18G+2/18H+2/18Y+2/18V
C	2/18R+2/18C+2/18G+2/18F+4/18S+2/18W+2/18Y+2/18STOP
E	2/18A+4/18D+2/18E+2/18Q+2/18G+2/18K+2/18V+2/18STOP
Q	2/18R+2/18E+2/18Q+4/18H+2/18L+2/18K+2/18P+2/18STOP
G	4/36A+6/36R+2/36D+2/36C+2/36E+12/36G+2/36S+1/36W+4/36V+1/36STOP
H	2/18R+2/18N+2/18D+4/18Q+2/18H+2/18L+2/18P+2/18Y
I	1/27R+2/27N+6/27I+4/27L+1/27K+3/27M+2/27F+2/27S+3/27T+3/27V
L	4/54R+2/54Q+2/54H+4/54I+18/54L+2/54M+6/54F+4/54P+2/54S+1/54W+6/54V+3/54STOP
K	2/18R+4/18N+2/18E+2/18Q+1/18I+2/18K+1/18M+2/18T+2/18STOP
M	1/9R+3/9I+2/9L+1/9K+1/9T+1/9V
F	2/18C+2/18I+6/18L+2/18F+2/18S+2/18Y+2/18V
P	4/36A+4/36R+2/36Q+2/36H+4/36L+12/36P+4/36S+4/36T
S	4/54A+6/54R+2/54N+4/54C+2/54G+2/54I+4/54L+2/54F+4/54P+14/54S+6/54T+1/54W+2/54Y+3/54STOP
T	4/36A+2/36R+2/36N+3/36I+2/36K+1/36M+4/36P+6/36S+12/36T
W	2/9R+2/9C+1/9G+1/9L+1/9S+2/9STOP
Y	2/18N+2/18D+2/18C+2/18H+2/18F+2/18S+2/18Y+4/18STOP
V	4/36A+2/36D+2/36E+4/36G+3/36I+6/36L+1/36M+2/36F+12/36V
STOP	2/27R+2/27C+2/27E+2/27Q+1/27G+3/27L+2/27K+3/27S+2/27W+4/27Y+4/27STOP

A, alanine; R, arginine; N, asparagine; D, aspartic acid; C, cysteine; E, glutamic acid; Q, glutamine; G, glycine; H, histidine; I, isoleucine; L, leucine; K, lysine; M, methionine; F, phenylalanine; P, proline; S, serine; T, threonine; W, tryptophan; Y, tyrosine; V, valine.

8.3.1. Future Amino-Acid Composition Mutated from Threonine

Assuming that we have a single amino acid, say, threonine, based on Tables 8-3 and 8-4, we know that it has 4/36 chance of mutating to alanine, 2/36 chance of mutating to arginine, 2/36 chance of mutating to asparagine, and so on. Let us see what will happen for the threonine. According to the calculation in left-hand column in Table 8-5, this threonine will mutate to nothing because no mutated amino acid has the chance larger than 0.5.

However, if we have threonines more than one, such as five, there will be something happened. As seen in the right-hand column in Table 8-5, there will be mutated amino acids alanine, proline, serine and threonine appeared because their probability is larger than 0.5.

This is very suggestive, because we generally know the composition of a protein, with which we can calculate the kind of amino acids that will appear according to Table 8-4. Especially, we can know when the STOP signal will appear because Table 8-4 does show such a possibility, so we can explain why a mutation can sometimes truncate a protein [135-137].

Table 8-5. Chance of threonine mutating to other amino acids

Chance of a single threonine mutating to other amino acids	Chance of five threonines mutating to other amino acids
1 threonine × 4/36 = 0.1111 alanines	5 threonines × 4/36 = 0.5556 alanines
1 threonine × 2/36 = 0.0556 arginines	5 threonines × 2/36 = 0.2778 arginines
1 threonine × 2/36 = 0.0556 asparagines	5 threonines × 2/36 = 0.2778 asparagines
1 threonine × 3/36 = 0.0833 isoleucines	5 threonines × 3/36 = 0.4167 isoleucines
1 threonine × 2/36 = 0.0556 lysines	5 threonines × 2/36 = 0.2778 lysines
1 threonine × 1/36 = 0.0278 methionines	5 threonines × 1/36 = 0.1389 methionines
1 threonine × 4/36 = 0.1111 prolines	5 threonines × 4/36 = 0.5556 prolines
1 threonine × 6/36 = 0.1667 serines	5 threonines × 6/36 = 0.8333 serines
1 threonine × 12/36=0.3333 threonines	5 threonines × 12/36=1.6667 threonines

8.3.2. Future Amino-Acid Composition Mutated from an Imaginary Protein

To understand, we first imagine that we have a particular protein containing each kind of 20 amino acids with the same number for each kind, that is, we have an imaginary protein with a length of 20 amino acids and accidentally each kind of amino acids has only one amino acid. Second, we assume that each amino acid of our imaginary protein has an equal chance to mutate, in other words, we would expect all 20 amino acids to mutate. Of course, these two conditions are man-made to demonstrate the calculation.

In Table 8-6, we calculate how this imaginary protein will mutate. This calculating process does indeed appear exhausting, however it can be easily done with computer program according to the format of Table 8-6.

In our assumption, we have 20 kinds of amino acids, and each kind has an amino acid. So the composition for these 20 kinds of amino acids is 5% for each kind. The last row in Table 8-6 shows the new composition after each amino acid has mutated. As this composition is purely related to the future, we call it as the future amino-acid composition, which actually is a time-oriented measure because it directs forwards future [135-137].

Figure 8-1 shows the more visible comparison between current and future amino-acid composition in this imaginary protein. We can see that there is no STOP signal in the current amino-acid composition, but the STOP signal appears in the future amino-acid composition. This is also an important point that our approach is different from others, which cannot estimate the STOP signal.

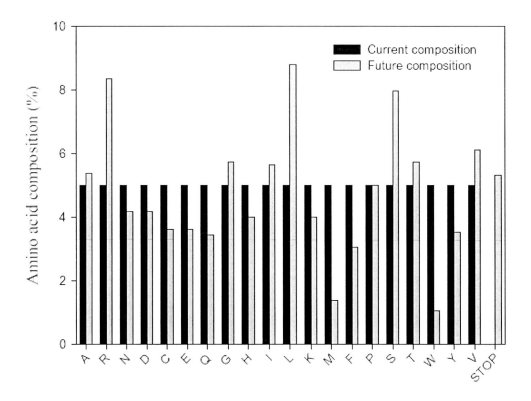

Figure 8-1. Current and future amino-acid compositions in an imaginary protein containing equal number of 20 kinds of amino acids.

8.3.3. Future Amino-Acid Composition Mutated from Hemoglobin β-Chain

We can imagine from the above example that a protein will have different future amino-acid composition if its current amino-acid composition is different from that of our imaginary protein in Table 8-6. Let us calculate another example, i.e. we will calculate the future amino-acid composition in human hemoglobin β-chain.

In Chapter 6, we have analyzed the actual and predicted distribution probabilities in human hemoglobin β-chain (Table 6-8). Here we show its amino-acid composition in Table 8-7.

Table 8-6. Calculation of future amino-acid composition in an imaginary protein

	A	R	N	D	C	E	Q	G	H	I
A×	12/36=0.3333			2/36=0.0556		2/36=0.0556		4/36=0.1111		1/54=0.0185
R×		18/54=0.3333		2/18=0.1111	2/54=0.0370		2/54=0.0370	6/54=0.1111	2/54=0.0370	2/18=0.1111
N×		2/18=0.1111	2/18=0.1111	2/18=0.1111					2/18=0.1111	
D×	2/18=0.1111	2/18=0.1111	2/18=0.1111		2/18=0.1111	4/18=0.2222		2/18=0.1111	2/18=0.1111	
C×								2/18=0.1111		
E×	2/18=0.1111			4/18=0.2222		2/18=0.1111	2/18=0.1111	2/18=0.1111		
Q×						2/18=0.1111	2/18=0.1111		4/18=0.2222	
G×	4/36=0.1111	6/36=0.1667		2/36=0.0556	2/36=0.0556	2/36=0.0556		12/36=0.3333		
H×		2/18=0.1111	2/18=0.1111	2/18=0.1111			4/18=0.2222		2/18=0.1111	
I×		1/27=0.0370	2/27=0.0741							6/27=0.2222
L×		4/54=0.0741					2/54=0.0370		2/54=0.0370	4/54=0.0741
K×		2/18=0.1111	4/18=0.2222			2/18=0.1111	2/18=0.1111			1/18=0.0556
M×		1/9=0.1111								3/9=0.3333
F×					2/18=0.1111					2/18=0.1111
P×	4/36=0.1111	4/36=0.1111					2/36=0.0556		2/36=0.0556	
S×	4/54=0.0741	6/54=0.1111	2/54=0.0370		4/54=0.0741			2/54=0.0370		2/54=0.0370
T×	4/36=0.1111	2/36=0.0556	2/36=0.0556							3/36=0.0833
W×		2/9=0.2222			2/9=0.2222			1/9=0.1111		
Y×			2/18=0.1111	2/18=0.1111	2/18=0.1111				2/18=0.1111	
V×	4/36=0.1111			2/36=0.0556		2/36=0.0556		4/36=0.1111		3/36=0.0833
	A	R	N	D	C	E	Q	G	H	I
Total	1.0741	1.6667	0.8333	0.8333	0.7222	0.7222	0.6852	1.1481	0.7963	1.1296
%	5.3704	8.3333	4.1667	4.1667	3.6111	3.6111	3.4259	5.7407	3.9815	5.6481

Table 8-6. Continued

	L	K	M	F	P	S	T	W	Y	V	STOP
A×					4/36=0.1111	4/36=0.1111	4/36=0.1111			4/36=0.1111	
R×	4/54=0.0741	2/54=0.0370	1/54=0.0185		4/54=0.0741	6/54=0.1111	2/54=0.0370	2/54=0.0370	2/18=0.1111		2/54=0.0370
N×		4/18=0.2222				2/18=0.1111	2/18=0.1111				
D×									2/18=0.1111	2/18=0.1111	
C×				2/18=0.1111		4/18=0.2222		2/18=0.1111	2/18=0.1111		2/18=0.1111
E×	2/18=0.1111	2/18=0.1111								2/18=0.1111	2/18=0.1111
Q×		2/18=0.1111			2/18=0.1111						2/18=0.1111
G×						2/36=0.0556		1/36=0.0278		4/36=0.1111	1/36=0.0278
H×	2/18=0.1111				2/18=0.1111				2/18=0.1111		
I×	4/27=0.1481	1/27=0.0370	3/27=0.1111	2/27=0.0741		2/27=0.0741	3/27=0.1111			3/27=0.1111	
L×	18/54=0.3333	2/54=0.0370	2/54=0.0370	6/54=0.1111	4/54=0.0741	2/54=0.0370		1/54=0.0185		6/54=0.1111	3/54=0.0556
K×	2/9=0.2222	2/18=0.1111	1/18=0.0556				2/18=0.1111			1/9=0.1111	
M×	6/18=0.3333	1/9=0.1111					1/9=0.1111				
F×				2/18=0.1111		2/18=0.1111			2/18=0.1111	2/18=0.1111	2/18=0.1111
P×					12/36=0.3333	4/36=0.1111	4/36=0.1111				
S×		2/36=0.0556		2/54=0.0370	4/54=0.0741	14/54=0.2593	6/54=0.1111	1/54=0.0185	2/54=0.0370		3/54=0.0556
T×			1/36=0.0278		4/36=0.1111	6/36=0.1667	12/36=0.3333				
W×	1/9=0.1111					1/9=0.1111					2/9=0.2222
Y×				2/18=0.1111		2/18=0.1111			2/18=0.1111		4/18=0.2222
V×	6/36=0.1667		1/36=0.0278	2/36=0.0556						12/36=0.3333	
Total	1.7593	0.7963	0.2778	0.6111	1	1.5926	1.1481	0.2130	0.7037	1.2222	1.0648
%	8.7963	3.9815	1.3889	3.0556	5	7.9630	5.7407	1.0648	3.5185	6.1111	5.3241

	L	K	M	F	P	S	T	W	Y	V	STOP
	L	K	M	F	P	S	T	W	Y	V	STOP

Table 8-7. Amino-acid composition of human hemoglobin β–chain

Amino acid	Number	Composition (%)
Alanine, A	15	10.2041
Arginine, R	3	2.0408
Asparagine, N	6	4.0816
Aspartic acid, D	7	4.7619
Cysteine, C	2	1.3605
Glutamic acid, E	8	5.4422
Glutamine, Q	3	2.0408
Glycine, G	13	8.8435
Histidine, H	9	6.1224
Isoleucine, I	0	0
Leucine, L	18	12.2449
Lysine, K	11	7.4830
Methionine, M	2	1.3605
Phenylalanine, F	8	5.4422
Proline, P	7	4.7619
Serine, S	5	3.4014
Threonine, T	7	4.7619
Tryptophan, W	2	1.3605
Tyrosine, Y	3	2.0408
Valine, V	18	12.2449
Total	147	100

We can use the current amino-acid composition in Table 8-7 to calculate the future amino-acid composition of human hemoglobin β-chain. Table 8-8 once again shows this exhausting calculation process, which can be easily implemented into computer programs. On the other hand, we should make efforts to visualize these results as people say that a picture is worthy 1000 words, otherwise we should mention the table containing different numbers. Figure 8-2 illustrates this visualization, which in fact combines the amino-acid compositions in both Tables 8-7 and 8-8.

Comparing Figure 8-1 with Figure 8-2, we can see the difference between our imaginary protein and human hemoglobin β-chain. Actually both figures implicate the time-oriented randomness, which not only forces the mutation, but more importantly indicates which kind of amino acid has a larger chance of appearing after mutation and which kind of amino acid has a smaller chance of appearing after mutation [135-137].

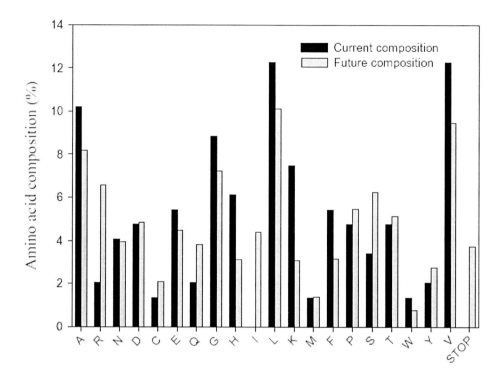

Figure 8-2. Current and future amino-acid compositions in human hemoglobin β-chain.

8.4. Future Mutated Amino-Acid Composition

One may have noticed that the amino-acid mutating probability in Table 8-4 contains the self-oriented mutation, for example, alanine (A) has 12/36 chance of mutating to alanine (A). One may wonder why Table 8-4 still contains the self-oriented mutation, as we have claimed the exclusion of self-oriented mutation in Section 8.2.

In fact, what we talk about exclusion of self-oriented mutation in Section 8.2 is referring to the RNA codon, that is, we do not count the point mutation in a RNA codon resulting in the same RNA codon. On the other hand, the self-oriented mutations in Table 8-4 result from the point mutation, which leads the RNA codon to mutate to different RNA codon that, however, again bring about the same amino acid as the original one.

It is no problem for us to completely exclude the self-oriented mutation at amino-acid level in Table 8-4, where however the RNA codon is changed anyway. This is an important reason that we need to extend our methods to the RNA codon level in Chapter 12.

Table 8-8. Calculation of future amino-acid composition in human hemoglobin β-chain

	A	R	N	D	C	E	Q	G	H	I
15A×	12/36=5									
3R×		18/54=1		2/36=0.8333	2/54=0.1111	2/36=0.8333	2/54=0.1111	4/36=1.6667	2/54=0.1111	1/54=0.0556
6N×		2/18=0.6667	2/18=0.6667					6/54=0.3333	2/18=0.6667	2/18=0.6667
7D×	2/18=0.7778	2/18=0.7778	2/18=0.7778	2/18=0.7778	2/18=0.2222	4/18=1.5556		2/18=0.7778	2/18=0.7778	
2C×					2/18=0.2222			2/18=0.2222		
8E×	2/18=0.8889	2/18=0.3333		4/18=1.7778		2/18=0.8889	2/18=0.8889	2/18=0.8889		
3Q×						2/18=0.3333	2/18=0.3333		4/18=0.6667	
13G×	4/36=1.4444	6/36=2.1667		2/36=0.7222	2/36=0.7222	2/36=0.7222		12/36=4.3333		
9H×		2/18=1	2/18=1	2/18=1			4/18=2		2/18=1	
0I×		1/27=0	2/27=0							6/27=0
18L×		4/54=1.3333					2/54=0.6667		2/54=0.6667	4/54=1.3333
11K×		2/18=1.2222	4/18=2.4444			2/18=1.2222	2/18=1.2222			1/18=0.6111
2M×		1/9=0.2222								3/9=0.6667
8F×					2/18=0.8889					2/18=0.8889
7P×	4/36=0.7778	4/36=0.7778					2/36=0.3889	2/36=0.3889	2/36=0.3889	
5S×	4/54=0.3704	6/54=0.5556	2/54=0.1852		4/54=0.3704			2/54=0.1852		2/54=0.1852
7T×	4/36=0.7778	2/36=0.3889	2/36=0.3889							3/36=0.5833
2W×		2/9=0.4444			2/9=0.4444			1/9=0.2222		
3Y×			2/18=0.3333	2/18=0.3333	2/18=0.3333				2/18=0.3333	
18V×	4/36=2		2/18=0.3333	2/36=1		2/36=1		4/36=2	2/18=0.3333	3/36=1.5000
	A	R	N	D	C	E	Q	G	H	I
Total	12.0370	9.6667	5.7963	7.1111	3.0926	6.5556	5.6111	10.6296	4.6111	6.4907
%	8.1885	6.5760	3.9431	4.8375	2.1038	4.4596	3.8171	7.2310	3.1368	4.4155

Table 8-8. Continued

	L	K	M	F	P	S	T	W	Y	V	STOP
15A×					4/36=1.6667	4/36=1.6567	4/36=1.6667			4/36=1.6667	
3R×	4/54=0.2222	2/54=0.1111	1/54=0.0556		4/54=0.2222	6/54=0.3333	2/54=0.1111	2/54=0.1111			2/54=0.1111
6N×	4/18=1.3333	4/18=1.3333				2/18=0.6667	2/18=0.6667		2/18=0.6667		
7D×									2/18=0.7778	2/18=0.7778	
2C×				2/18=0.2222		4/18=0.4444		2/18=0.2222	2/18=0.2222		2/18=0.2222
8E×		2/18=0.8889								2/18=0.8889	2/18=0.8889
3Q×	2/18=0.3333	2/18=0.3333			2/18=0.3333						2/18=0.3333
13G×						2/36=0.7222		1/36=0.3611		4/36=1.4444	1/36=0.3611
9H×	2/18=1				2/18=1				2/18=1		
0I×	4/27=0	1/27=0	3/27=0	2/27=0		2/27=0	3/27=0			3/27=0	
18L×	18/54=6		2/54=0.6667	6/54=2	4/54=1.3333	2/54=0.6667		1/54=0.3333		6/54=2	3/54=1
11K×	2/18=1.2222	2/18=1.2222	1/18=0.6111				2/18=1.2222				2/18=1.2222
2M×	2/9=0.4444	1/9=0.2222					1/9=0.2222			1/9=0.2222	
8F×	6/18=2.6667			2/18=0.8889		2/18=0.8889			2/18=0.8889	2/18=0.8889	
7P×	4/36=0.7778					4/36=0.7778	4/36=0.7778				
5S×	2/54=0.1852	2/36=0.3889	1/36=0.1944	2/54=0.1852	4/54=0.3704	14/54=1.2963	6/54=0.5556	1/54=0.0926	2/54=0.1852		3/54=0.2778
7T×					4/36=0.7778	6/36=1.1667	12/36=2.3333				
2W×	1/9=0.2222					1/9=0.2222					2/9=0.4444
3Y×				2/18=0.3333		2/18=0.3333			2/18=0.3333		4/18=0.6667
18V×	6/36=3		1/36=0.5	2/36=1						12/36=6	
	L	K	M	F	P	S	T	W	Y	V	STOP
Total	14.8519	4.5000	2.0278	4.6296	8.0370	9.1852	7.5556	1.1204	4.0741	13.8889	5.5278
%	10.1033	3.0612	1.3794	3.1494	5.4674	6.2484	5.1398	0.7622	2.7715	9.4482	3.7604

After excluding these self-oriented mutations from Table 8-4, we can get the amino-acid mutating probability without self-oriented mutations. The process for this exclusion is as follows. We simply subtract the self-oriented mutations from Table 8-4. For example, alanine "A" has a total of 36 options of mutation, of which 12 are alanines. If we subtract 12 options from 36 options, we have a total of 24 options for the rest mutated amino acids. This way, we can get the amino-acid mutating probability excluding self-oriented mutation at amino-acid level in Table 8-9.

At early stage of developing the measure for time-oriented randomness [135-137], we were more interested in the amino-acid mutating probability excluding self-oriented mutations in Table 8-9 than in that including self-oriented mutations in Table 8-4, because we considered that the amino-acid mutating probability excluding self-oriented mutation is more suitable for the prediction of mutation trends, which in fact is true. However, we soon found that Table 8-9 is mainly suitable for the comparison among different proteins rather than the estimation of mutation probability in each amino acid in a protein in question. Thus our recent studies are more related to the amino-acid mutating probability including self-oriented mutations in Table 8-4.

8.4.1. Future Mutated Amino-Acid Composition from an Imaginary Protein

The analysis in this section is much similar to the analysis in Section 8.3.2, where we have an imaginary protein containing all 20 kinds of amino acids with the same number for each kind and then we detail the calculations of future amino-acid composition through Table 8-6. In this section, we will conduct the similar calculations although the process is boring and terrific, however it would be helpful for the readers who skim this book.

Table 8-10 once again shows the detailed calculation process. Compared with the calculations in Table 8-6, we can see that there is no difference in only methionine "M" and tryptophan "W". Figure 8-3 displays the future mutated amino-acid composition for this imaginary protein in comparison with its current and future amino-acid compositions. This figure indicates that we would expect to see more mutated arginine "R", leucine "L" and serine "S", but few methionine "M" and tryptophan "W". Due to the subtraction of self-oriented mutations, there would be a larger chance of occurring STOP in future.

8.4.2 Future Mutated Amino-Acid Composition from Hemoglobin β-Chain

In this section, we only briefly show the future mutated amino-acid composition from human hemoglobin β-chain based on its current amino-acid composition in Table 8-7. Figure 8-4 illustrates the future mutated amino-acid composition from hemoglobin β-chain in comparison with its current and future amino-acid compositions. Figure 8-4 does not tell us more things de facto if we do not compare the future mutated amino-acid composition among proteins of interests.

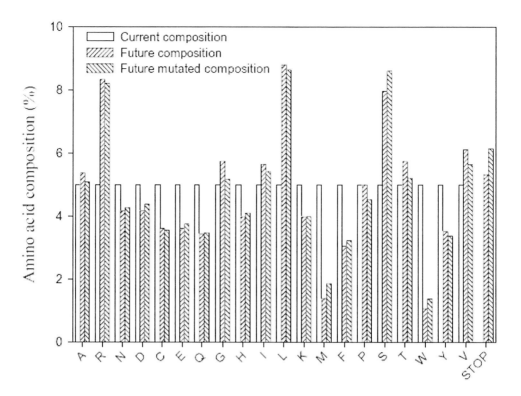

Figure 8-3. Current, future and future mutated amino-acid compositions from an imaginary protein.

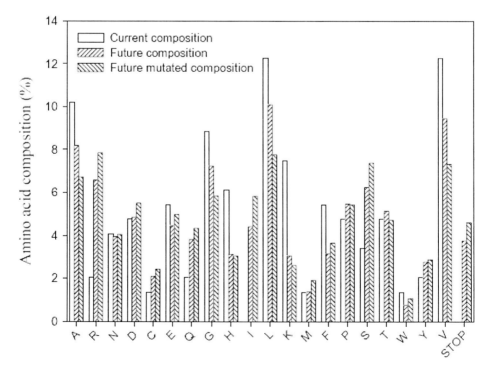

Figure 8-4. Current, future and future mutated amino-acid compositions in human hemoglobin β-chain.

Table 8-9. Amino-acid mutating probability excluding self-oriented mutation at amino-acid level

Original amino acid	Mutated amino acids with their probability
A	2/24D+2/24E+4/24G+4/24P+4/24S+4/24T+4/24V
R	2/36C+6/36G+2/36H+1/36I+2/36K+4/36L+1/36M+4/36P+2/36Q+6/36S+2/36T+2/36W+2/36STOP
N	2/16D+2/16H+2/16I+4/16K+2/16S+2/16T+2/16Y
D	2/16A+4/16E+2/16G+2/16H+2/16N+2/16V+2/16Y
C	2/16F+2/16G+2/16R+4/16S+2/16W+2/16Y+2/16STOP
E	2/16A+4/16D+2/16G+2/16K+2/16Q+2/16V+2/16STOP
Q	2/16E+4/16H+2/16K+2/16L+2/16P+2/16R+2/16STOP
G	4/24A+2/24C+2/24D+2/24E+6/24R+2/24S+4/24V+1/24W+1/24STOP
H	2/16D+2/16L+2/16N+2/16P+4/16Q+2/16R+2/16Y
I	2/21F+1/21K+4/21L+3/21M+2/21N+1/21R+2/21S+3/21T+3/21V
L	6/36F+2/36H+4/36I+2M/36+4/36P+2/36Q+4/36R+2/36S+6/36V+1/36W+3/36STOP
K	2/16E+1/16I+1/16M+4/16N+2/16Q+2/16R+2/16T+2/16STOP
M	3/9I+1/9K+2/9L+1/9R+1/9T+1/9V
F	2/16C+2/16I+6/16L+2/16S+2/16V+2/16Y
P	4/24A+2/24H+4/24L+2/24Q+4/24R+4/24S+4/24T
S	4/40A+4/40C+2/40F+2/40G+2/40I+2/40L+2/40N+4/40?+6/40R+6/40T+1/40W+2/40Y+3/40STOPS
T	4/24A+2/24R+2/24N+3/24I+2/24K+1/24M+4/24P+6/24S
W	2/9C+1/9G+1/9L+2/9R+1/9S+2/9STOP
Y	2/16C+2/16D+2/16F+2/16H+2/16N+2/16S+4/16STOP
V	4/24A+2/24D+2/24E+2/24F+4/24G+3/24I+6/24L+/24M
STOP	2/23C+2/23E+1/23G+2/23K+3/23L+2/23Q+2/23R+3/23S+2/23W+4/23Y

Table 8-10. Calculation of future mutated amino-acid composition in an imaginary protein

	A	R	N	D	C	E	Q	G	H	I
A×				2/24=0.0833		2/24=0.0833		4/24=0.1667		
R×					2/36=0.0556		2/36=0.0556	6/36=0.1667	2/36=0.0556	1/36=0.0278
N×			2/16=0.1250	2/16=0.1250					2/16=0.1250	2/16=0.1250
D×	2/16=0.1250	2/16=0.1250				4/16=0.2500		2/16=0.1250	2/16=0.1250	
C×	2/16=0.1250							2/16=0.1250		
E×	2/16=0.1250	2/16=0.1250		4/16=0.2500			2/16=0.1250	2/16=0.1250		
Q×			2/16=0.1250			2/16=0.1250				
G×	4/24=0.1667	6/24=0.2500		2/24=0.0833	2/24=0.0833	2/24=0.0883			4/16=0.2500	
H×		2/16=0.1250		2/16=0.1250			4/16=0.2500			
I×		1/21=0.0476	2/21=0.0952							
L×		4/36=0.1111					2/36=0.0556		2/36=0.0556	4/36=0.1111
K×		2/16=0.1250	4/16=0.2500			2/16=0.1250	2/16=0.1250			1/16=0.0625
M×		1/9=0.1111								3/9=0.3333
F×					2/16=0.1250					2/16=0.1250
P×	4/24=0.1667	4/24=0.1667					2/24=0.0833		2/24=0.0833	
S×	4/40=0.1000	6/40=0.1500	2/40=0.0500		4/40=0.1000			2/40=0.0500		2/40=0.0500
T×	4/24=0.1667	2/24=0.0833	2/24=0.0833							3/24=0.1250
W×		2/9=0.2222			2/9=0.2222			1/9=0.1111		
Y×			2/16=0.1250	2/16=0.1250	2/16=0.1250				2/16=0.1250	
V×	4/24=0.1667			2/24=0.0833		2/24=0.0833		4/24=0.1667		3/24=0.1250
	A	R	N	D	C	E	Q	G	H	I
Total	1.0167	1.6421	0.8536	0.8750	0.7111	0.7500	0.6944	1.0361	0.8194	1.0847
%	5.0833	8.2103	4.2679	4.3750	3.5556	3.7500	3.4722	5.1806	4.0972	5.4236

	L	K	M	F	P	S	T	W	Y	V	STOP
A×					4/24=0.1667	4/24=0.1667	4/24=0.1667			4/24=0.1667	
R×	4/36=0.1111	2/36=0.0556	1/36=0.0278		4/36=0.1111	6/36=0.1667	2/36=0.0556	2/36=0.0556			2/36=0.0556
N×		4/16=0.2500				2/16=0.1250	2/16=0.1250				
D×									2/16=0.1250	2/16=0.1250	
C×				2/16=0.1250		4/16=0.2500		2/16=0.1250	2/16=0.1250		2/16=0.1250
E×		2/16=0.1250							2/16=0.1250	2/16=0.1250	2/16=0.1250
Q×	2/16=0.1250	2/16=0.1250			2/16=0.1250						2/16=0.1250
G×						2/24=0.0833		1/24=0.0417		4/24=0.1667	1/24=0.0417
H×	2/16=0.1250				2/16=0.1250				2/16=0.1250		
I×	4/21=0.1905	1/21=0.0476	3/21=0.1429	2/21=0.0952		2/21=0.0952	3/21=0.1429			3/21=0.1429	
L×			2/36=0.0556	6/36=0.1667	4/36=0.1111	2/36=0.0556		1/36=0.0278		6/36=0.1667	3/36=0.0833
K×			1/16=0.0625				2/16=0.1250				2/16=0.1250
M×	2/9=0.2222	1/9=0.1111					1/9=0.1111			1/9=0.1111	
F×	6/16=0.3750					2/16=0.1250			2/16=0.1250	2/16=0.1250	
P×	4/24=0.1667				4/24=0.1667	4/24=0.1667	4/24=0.1667				
S×	2/40=0.0500			2/40=0.0500	4/40=0.1000	6/24=0.2500	6/40=0.1500	1/40=0.0250	2/40=0.0500		3/40=0.0750
T×		2/24=0.0833	1/24=0.0417		4/24=0.1667						
W×	1/9=0.1111					1/9=0.1111					2/9=0.2222
Y×				2/16=0.1250		2/16=0.1250					4/16=0.2500
V×	6/24=0.2500		1/24=0.0417	2/24=0.0833							
	L	K	M	F	P	S	T	W	Y	V	STOP
Total	1.7266	0.7976	0.3720	0.6452	0.9056	1.7202	1.0429	0.2750	0.6750	1.1290	1.2278
%	8.6329	3.9881	1.8601	3.2262	4.5278	8.6012	5.2143	1.3750	3.3750	5.6448	6.1389

8.4.3. Future Mutated Amino-Acid Composition Among Several Proteins

When comparing the future mutated amino-acid composition with the current one, we in fact can find the mutation trend with respect to each kind of amino acids. Among three proteins presented in Figure 8-5, for example, we can find that many glutamic acids "E" will disappear after mutation. Moreover, some kinds of amino acids will decrease their number after mutation, such as isoleucine "I" and leucine "L" in goose H5N1 matrix protein 2 (the upper panel), lysine "K" and tyrosine "Y" in human Bruton's tyrosine kinase (the middle panel), and asparagine "N", lysine "K" and valine "V" in *Escherichia coli* Pilx9 protein (the lower panel).

On the other hand, some kinds of amino acids will appear after mutation. For instance, alanine "A" and arginine "R" will increase in both goose H5N1 matrix protein 2 and human Bruton's tyrosine kinase. Cysteine "C" and histidine "H" will increase in *Escherichia coli* Pilx9 protein. Also, the STOP signals will appear more in three proteins.

In Table 8-11, we can see that human Bruton's tyrosine kinase has the smallest sum of absolute difference (26.9 and 29.6), by contrast, *Escherichia coli* Pilx9 protein has the largest sum of absolute difference (29.4 and 33.3). These results suggest that we would expect to see fewer mutations in human Bruton's tyrosine kinase than in *Escherichia coli* Pilx9 protein, while the number of mutations in goose H5N1 matrix protein 2 would be between above two proteins. Of course, this implication is based on the assumption that the biological world has a similar period of time to reach the final distribution patterns governed by the future amino-acid composition. In such a case, *Escherichia coli* Pilx9 protein would have a relative fast speed of evolution in remaining time.

Still, the smallest sum of absolute difference in human Bruton's tyrosine kinase may imply that human Bruton's tyrosine kinase is the most stable protein among three proteins in Table 8-11.

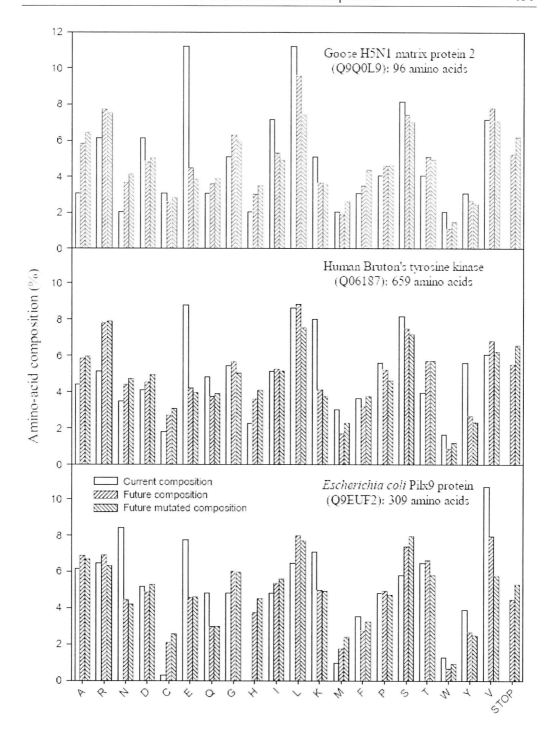

Figure 8-5. Current, future and future mutated amino-acid compositions in three proteins.

Table 8-11. Comparison of relative speed of evolution among three proteins presented in Figure 8-5

	Goose H5N1 matrix protein 2					Human Bruton's tyrosine kinase					Escherichia coli Pilx9 protein				
	I	II	III	II-I	III-I	I	II	III	II-I	III-I	I	II	III	II-I	III-I
A	3.1	5.8	6.5	2.8	3.4	4.4	5.8	5.9	1.4	1.6	6.1	6.9	6.7	0.8	0.5
R	6.1	7.7	7.5	1.6	1.4	5.2	7.8	7.9	2.6	2.7	6.5	6.9	6.3	0.5	0.1
N	2.0	3.7	4.1	1.6	2.1	3.5	4.4	4.7	0.9	1.3	8.4	4.4	4.2	3.9	4.2
D	6.1	4.8	5.0	1.3	1.1	4.1	4.6	5.0	0.5	0.9	5.2	4.9	5.3	0.3	0.1
C	3.1	2.6	2.8	0.5	0.2	1.8	2.7	3.1	0.9	1.3	0.3	2.1	2.6	1.8	2.2
E	11.2	4.5	3.9	6.7	7.4	8.8	4.2	4.0	4.6	4.8	7.7	4.6	4.6	3.2	3.1
Q	3.1	3.6	3.9	0.5	0.8	4.8	3.8	3.9	1.1	0.9	4.8	3.0	3.0	1.9	1.9
G	5.1	6.3	6.0	1.2	0.9	5.5	5.7	5.0	0.2	0.4	4.8	6.0	6.0	1.2	1.1
H	2.0	3.0	3.5	1.0	1.5	2.3	3.6	4.1	1.3	1.8	0.0	3.8	4.5	3.8	4.5
I	7.1	5.3	4.9	1.8	2.2	5.2	5.3	5.2	0.1	0.0	4.8	5.4	5.6	0.5	0.8
L	11.2	9.6	7.5	1.6	3.8	8.6	8.8	7.5	0.2	1.1	6.5	8.0	7.7	1.5	1.3
K	5.1	3.7	3.6	1.4	1.5	8.0	4.1	3.8	3.9	4.3	7.1	5.0	4.9	2.1	2.2
M	2.0	1.9	2.6	0.1	0.6	3.0	1.7	2.3	1.3	0.7	1.0	1.8	2.4	0.8	1.4
F	3.1	3.5	4.4	0.4	1.3	3.6	3.2	3.8	0.4	0.1	3.5	2.7	3.2	0.8	0.3
P	4.1	4.6	4.6	0.5	0.5	5.6	5.2	4.6	0.4	1.0	4.8	4.9	4.7	0.1	0.1
S	8.2	7.4	7.0	0.7	1.1	8.2	7.5	7.2	0.7	1.0	5.8	7.4	8.0	1.6	2.2
T	4.1	5.1	4.9	1.0	0.9	3.9	5.7	5.7	1.8	1.8	6.5	6.6	5.8	0.2	0.6
W	2.0	1.1	1.5	0.9	0.6	1.7	0.9	1.2	0.8	0.5	1.3	0.7	0.9	0.6	0.4
Y	3.1	2.7	2.5	0.4	0.6	5.6	2.7	2.3	2.9	3.3	3.9	2.7	2.5	1.2	1.4
V	7.1	7.8	7.1	0.7	0.0	6.1	6.8	6.2	0.8	0.2	10.6	8.0	5.8	2.7	4.9
Sum				27.0	31.9				26.9	29.6				29.4	33.3

I, II and III indicate the current, future and future mutated amino-acid composition (%), respectively; II-I and III-I are in absolute values.

Chapter 9

Behavior of Future Amino-Acid Composition and its Application

As we have already done in Chapters 4 and 6, with which you are familiar, we will investigate the behavior of future amino-acid composition from spatial and time angles in order to know if this measure is alive. On the other hand, this measure is somewhat similar to the amino-acid distribution probability, which is relevant to a kind of amino acids rather than an individual amino acid.

In this chapter we will mix the behavior of future amino-acid composition with its applications, because the future amino-acid composition is based on the translation probability between RNA codons and translated amino acids, say, it is based on large-scale and long-term statistics.

As we toss a coin, we in principle know that each side of coin has a half a chance up and another half down in large-scale and long-term tossing, however we cannot predict the result of a particular tossing. Similarly, we cannot precisely predict which kind of amino acid would be formed after mutation. This is the first difficulty preventing us from using the sophisticated mathematical tools to deal with the future amino-acid composition.

Moreover, the amino-acid mutating probability in Table 8-4 indicates the relationship between original and mutated amino acids, but this relationship is also difficult to use deterministic mathematical tools to model. Therefore our application in this Chapter is less mathematically sophisticated.

9.1. Behavior of Future Amino-Acid Composition from Spatial Angle

The biggest difference between the future amino-acid composition and two previous measures, amino-acid pair predictability and amino-acid distribution probability, is that the future amino-acid composition is talking about future, say, after a series of mutations, although we want to compare its behavior from spatial angle. On the other hand, the behaviors of amino-acid pair predictability and amino-acid distribution probability is only relevant to the current situation when compare their behavior from spatial angle.

9.1.1. Future Amino-Acid Composition in Proteins with Different Lengths

The proteins with different lengths might have the similar amino-acid compositions, or more frequently have different amino-acid compositions. The computation in Chapter 8 has already detailed the calculating process from current to future amino-acid composition, such as Table 8-6. Thus, we would expect that the future amino-acid composition would be the same for proteins with the same amino-acid composition no matter how difference in their lengths.

Actually, the amino-acid composition can be very different from protein to protein. Figure 9-1 shows the amino-acid compositions of five proteins with very different lengths. We can easily draw the conclusion that the behavior of future amino-acid composition is subject to the current amino-acid composition. On the other hand, we can also see how the behavior of future amino-acid composition is different from the amino-acid pair predictability and amino-acid distribution probability, whose behavior is only related to one state, in plain words, we only need to use a single bar to present a kind of amino acid, but here we need to use two bars to present a kind of amino acid because there are two states, the current and future.

Considering these two states, we can see that the difference between the least and most abundant amino acids is larger in the current state than that in the future one, say, the difference between black bars is larger than that between gray ones. This is very suggestive because it implies that the mutations have the trend to reduce the difference between different kinds of amino acids. To recap, we would say that the amino-acid mutating probability in Table 8-4 is based on the random mechanism because it is obtained from statistics between RNA codons and translated amino acids. This further supports our previous argument that randomness engineers mutations [39, 63, 64], as the difference of amino-acid composition decreases through mutations [136-137].

9.1.2. Future Amino-Acid Composition in Proteins with Similar Length

Herein we take three human proteins as example to analyze their current and future amino-acid compositions (Table 9-1). They are the collagen $\alpha1(I)$ chain precursor (CA11, accession number P02452), collagen $\alpha1(III)$ chain precursor (CA13, accession number P02461) and copper-transporting ATPase 2 (AT7B, accession number P35670). They contain 1464, 1466 and 1465 amino acids, respectively, so that their length is quite similar.

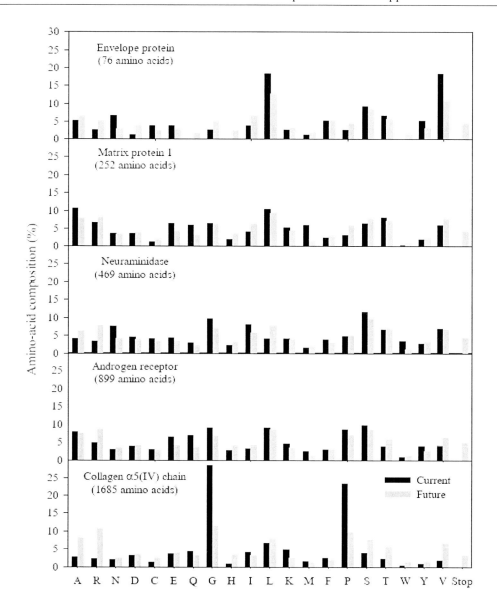

Figure 9-1. Current and future amino-acid compositions in 5 proteins with different lengths.

Table 9-1. Current and future amino-acid compositions in 3 proteins with similar length

Amino acid	Current composition (%)			Future composition (%)			Future composition (%) / Current composition (%)		
	CA11	CA13	AT7B	CA11	CA13	AT7B	CA11	CA13	AT7B
A	9.63	7.84	9.15	10.34	9.72	7.78	1.07	1.24	0.85
R	4.85	4.09	3.62	10.30	10.49	7.22	2.12	2.56	1.99
N	1.91	2.80	3.34	2.20	2.37	3.60	1.15	0.85	1.08
D	4.51	3.75	4.23	4.21	4.20	3.94	0.93	1.12	0.93
C	1.23	1.50	1.98	2.50	2.64	2.01	2.03	1.76	1.02
E	5.12	5.05	5.46	4.57	4.33	4.08	0.89	0.86	0.75

Table 9-1. Continued

Amino acid	Current composition (%)			Future composition (%)			Future composition (%) / Current composition (%)		
	CA11	CA13	AT7B	CA11	CA13	AT7B	CA11	CA13	AT7B
Q	3.28	2.93	4.71	2.87	2.91	2.99	0.88	0.99	0.63
G	26.71	28.17	7.10	12.26	12.36	6.55	0.46	0.44	0.92
H	0.61	0.95	2.46	2.96	2.93	3.08	4.85	3.08	1.25
I	1.64	2.46	7.24	2.29	2.53	5.86	1.40	1.03	0.81
L	3.28	3.27	9.01	5.77	5.75	9.18	1.76	1.76	1.02
K	3.96	4.23	5.12	2.31	2.47	3.49	0.58	0.58	0.68
M	0.89	1.16	3.00	0.78	0.83	1.91	0.88	0.72	0.64
F	1.84	1.57	2.46	1.25	1.32	3.06	0.68	0.84	1.24
P	18.99	19.17	4.91	9.07	8.84	5.64	0.48	0.46	1.15
S	4.10	4.98	8.60	7.85	7.96	7.74	1.91	1.60	0.90
T	3.01	2.11	5.60	5.75	5.59	6.59	1.91	2.65	1.18
W	0.41	0.48	0.75	1.20	1.26	0.88	2.93	2.63	1.17
Y	0.89	1.02	1.64	1.39	1.49	2.12	1.56	1.46	1.29
V	3.14	2.46	9.62	7.00	6.73	8.50	2.23	2.74	0.88
STOP				3.14	3.26	3.77			

CA11, collagen □1(I) chain precursor (Accession number P02452); CA13, collagen □1(III) chain precursor (Accession number P02461); AT7B, copper-transporting ATPase 2 (Accession number P35670).

As can be seen in Table 9-1, the current amino-acid composition is remarkably different although these three proteins have very similar length. Naturally, their future amino-acid composition is different too. Still, we can calculate the ratio of future versus current amino-acid composition if we want to know the magnitude from the current state to its future one.

In the right three columns of Table 9-1, we can see that only lysine "K" has the same magnitude between CA11 and CA13 proteins (0.58 versus 0.58) although these two proteins have different numbers of lysines.

Furthermore, the ratio of future versus current amino-acid composition in fact indicates the mutation trend, that is, the bigger the ratio is, the larger the mutation trend is for a given kind of amino acids. From this ratio, we can know which kind of amino acids is more likely to mutate, which is very important for knowing the mutation trend in each kind of amino acids and predicting mutations, thus we can use this ratio as an indicator that randomness engineers mutation.

9.1.3. Future Amino-Acid Composition in Different Subtypes

In this section, we will look at the behavior of future amino-acid composition in proteins from different subtypes in order to observe if this measure is different among subtypes. We can also analyze the mutation trend of hemagglutinins among these subtypes of influenza A viruses, as we have seen in the last section that the future amino-acid composition can indicate the mutation trend [135-137].

Figure 9-2 displays the current and future amino-acid compositions in hemagglutinins from various subtypes of influenza A viruses, which is somewhat research-oriented. For knowing the amount of amino acids, we use the number of amino acids rather than their percentage in this figure, by which we can exactly compare how many amino acids different between the current and future composition.

We can see that the differences between current and future amino-acid composition vary among different subtypes, even among 20 kinds of amino acids within the same subtype. This implies that the force of randomness engineers mutations. For example, asparagines "N" and glutamic acids "E" in H5 hemagglutinins have larger difference between current and future composition among 20 kinds of amino acids, thus we would expect to see more mutations occur in asparagines and glutamic acids then in other kinds of amino acids.

Considering different subtypes, the future amino-acid compositions are less different than the current ones, suggesting that future mutations will create new hemagglutinins more similar. This is the base for new subtype appearing.

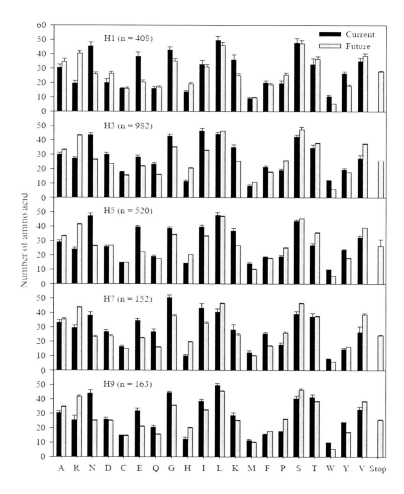

Figure 9-2. Current and future amino-acid compositions of hemagglutinins from different influenza A virus subtypes.

Another important point is the STOP signal that is located at the end of protein sequence and generally excluded from the calculation of amino-acid composition. However, the STOP signal can be mutated from some types of amino acids as the evolution process goes on [135-137], so that some mutations will truncate the hemagglutinins. Although there are some truncated proteins documented in databank, their real number would be larger because the truncated proteins generally are dysfunction leading to protein death.

9.1.4. Future Amino-Acid Composition in Proteins Cross Species

Similarly, we can also observe the behavior of future amino-acid composition in a protein family cross species. Figure 9-3 illustrates the current and future amino-acid compositions of H3 hemagglutinins among different species, of which human H3 hemagglutinins have been recorded in great details [138-140]. Moreover, the accumulation of amino-acid substitutions in the hemagglutinins promotes irreversible structural changes during evolution, and antigenic changes in H3 hemagglutinins may not be limited [141]. We can see what will happen in future.

First, the patterns of current and future amino-acid composition are not much different among species, which is the characteristic of the same protein family. On the other hand, this feature means that we can study the mutation patterns in some species to get the insight in other species.

Second, the standard deviation (SD) of future amino-acid composition is smaller than that of current one. This implies that mutations slightly converge in large-scale, so some mutant hemagglutinins will be less different between species, which can potentially link to the cross-species infection. For example, the accumulated evidence currently shows that avian H5N1 influenza A virus can infect human and other mammalian [65-67]. Accordingly, we would expect to see more cross-species infections in future.

Third, there are considerable STOP signals, which will appear in future as we have seen in Figures 9-1 and 9-2. Although we have mentioned that there will be shorter proteins formed in future, the more important implication for evolution is that the mutations will produce more and more different proteins because the position holding STOP signal is unpredictable. This may become one of the underlined reasons that more and more species appear during the evolution.

9.2 Behavior of Future Amino-Acid Composition from Time Angle

Although the behavior of future amino-acid composition in the above sections has the clear time direction toward future, this time direction is in fact ambiguous, that is, we do not know exactly when the future amino-acid composition will be reached from its current one, perhaps this process will take hundreds or thousands of years.

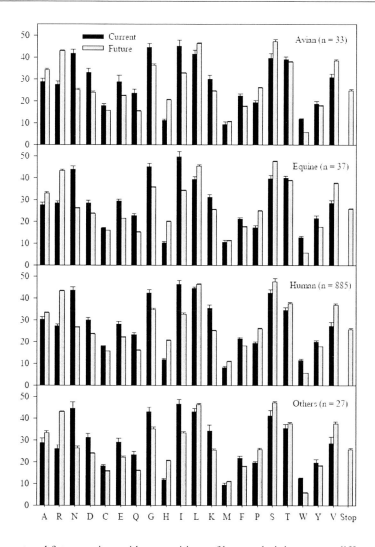

Figure 9-3. Current and future amino-acid compositions of hemagglutinins among different species of H3 influenza viruses.

If we would have a relatively clear time point on the formation of future amino-acid composition, we would jump from fitting the historical evolutionary process of proteins to simulate the future evolutionary process of proteins. With super-fast speed computation, this would be done rapidly, if we would understand more this process. More optimistically, we can predict the new proteins, new functions, even new species based on our understanding on all related issues.

All of these mean that we need at first to get more precise knowledge on the behavior of future amino-acid composition over time, that is, the behavior of future amino-acid composition along the time course. Figure 9-4 shows the current and future amino-acid compositions along the time course with respect to each kind of amino acids.

This figure is calculated based on 476 hemagglutinins of North America human H3N2 influenza viruses from 1968 to 2006. The calculation process is as the same as those

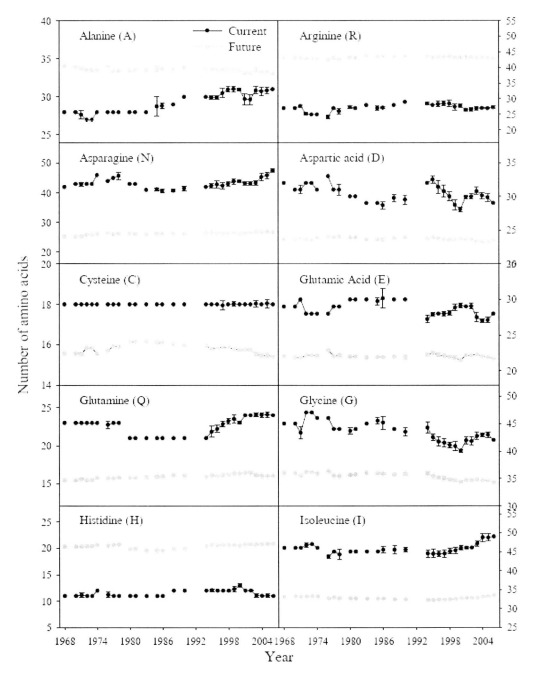

Figure 9-4A. Current and future amino-acid compositions along the time course. The data are presented as mean±SD (based on 476 hemagglutinins of North America human H3N2 influenza viruses).

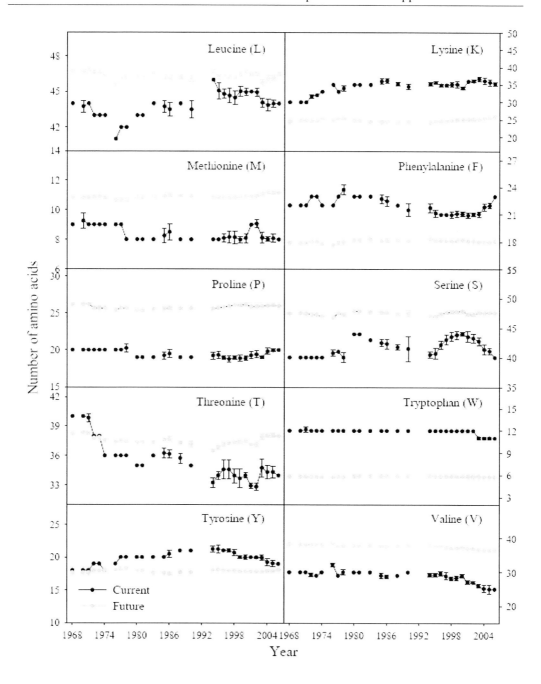

Figure 9-4B. Current and future amino-acid compositions along the time course. The data are presented as mean±SD (based on 476 hemagglutinins of North America human H3N2 influenza viruses).

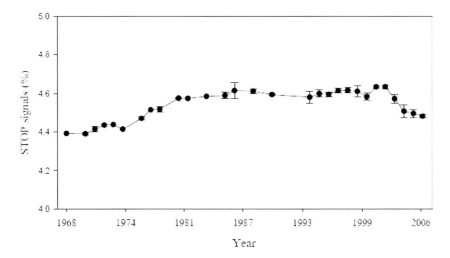

Figure 9-4C Future STOP signals along the time course. The data are presented as mean±SD (based on 476 hemagglutinins of North America human H3N2 influenza viruses).

presented in Tables 8-6 and 8-8. In fact, Figure 9-4 is more meaningful than Figure 3-5 because Figure 3-5 is a reliable record of history using our methods without future. Oppositely, Figure 9-4 has recorded the historical amino-acid compositions as well as the historically future ones, thus we can trace the changes in both compositions. For instance, alanines (A) generally follow their future composition, so that the difference between its current and future composition becomes narrowing.

Moreover, Figure 9-4C indicates that how many proteins can theoretically be truncated over time. Although we would expect to see only some of them to survive, they provide the resources for new proteins that may bring about new strains, subtypes, species and so on.

From the view of pure innovation in research, Figure 9-4 sets the best way for mathematical modeling because the differential equation is generally built as time function. We could imagine that we may build a system of differential equations for all 20 kinds of amino acids with STOP signal, which would pave the way for simulating the new proteins in future evolution with relatively exact time points. Nevertheless, this will be our future research topic.

9.3. Prediction of Would-be-Mutated Amino Acid

For the prediction of mutations, we need to predict which kind of amino acid will appear after mutation at the predicted position. In fact, Table 8-9 dictates the probability of would-be-mutated amino acids in a large-scale and long-term statistics.

However, the difficulty related to the large-scale and long-term statistics is that we cannot predict exactly what will happen in a single case as we stated at the beginning of this chapter.

On the other hand, we unfortunately could not define the force, which drives an amino acid to mutate to other one, thus we could not build any kinetic model to describe this process.

Perhaps the best way is to record all the historical mutated amino acids and their original amino acids, and then calculate their frequency and compare it with the corresponding frequency determined in Table 8-9. Accordingly, the bigger the difference between theoretical and historically accumulated frequency is, the larger the chance of occurrence of mutation is. Surely, the more precise the record is, the more precise the prediction is.

Figure 9-5 shows the graph along this line of thought. The upper panel is the amino-acid mutating probability according to Table 8-9, whose original and mutated amino acids contribute to the x- and y-axes and the amino-acid mutating probability contributes to the z-axis. There are 149 amino-acid mutating probabilities, so that some crosses between x- and y-axis have no probability, indicating that no mutated amino acid can theoretically be formed from the kind of original amino acid. The lower panel is the probability of mutated amino acids accumulated historically, which is obtained by dividing the number of each kind of mutated amino acids by the sum of all mutated amino acids from a certain kind of original amino acid. Table 9-2 lists several examples for the comparison.

Figure 9-5 shows the graph along this line of thought. The upper panel is the amino-acid mutating probability according to Table 8-9, whose original and mutated amino acids contribute to the x- and y-axes and the amino-acid mutating probability contributes to the z-axis. There are 149 amino-acid mutating probabilities, so that some crosses between x- and y-axis have no probability, indicating that no mutated amino acid can theoretically be formed from the kind of original amino acid. The lower panel is the probability of mutated amino acids accumulated historically, which is obtained by dividing the number of each kind of mutated amino acids by the sum of all mutated amino acids from a certain kind of original amino acid. Table 9-2 lists several examples for the comparison.

As seen in Figure 9-5, there are some differences in corresponding bars between the upper and lower panels. According to Table 8-9, the amino-acid mutating probability governs the large-scale and long-term mutations. In such a case, we would expect to see the difference between upper and lower panel diminishing along the time course. This is true in some cases as shown in columns 2 and 3 in Table 9-2, where the accumulated probability is as the same as or similar to its theoretical probability.

More meaningfully, Figure 9-5 actually told us the history, current state and future of protein family we are interested. For example, some mutations are recorded frequently, resulting in the accumulated probability bigger than its theoretical one in columns 4 and 5 in Table 9-2. By contrast, some mutations are documented rare, leading to the accumulated probability smaller than its theoretical one in columns 6 and 7 in Table 9-2.

The big difference for a particular amino acid between its theoretical and accumulated probability suggests that there must be some importantly underlying reason for the difference, which will indicate the direction for conducting research to trace the mutation causes for this particular amino acid. Still, the difference implies the survival rate of mutant proteins, whose functions should be directed toward the evolution. In return, these two lines of research, mutation causes and survival rate, will render new insight to the future.

Figure 9-6 illustrates the prediction of would-be-mutated amino acids in a protein, to which we can use either the mutating probability in Table 8-9 or the probability difference between upper and lower panel in Figure 9-5. There are four valines "V" at positions 220,

Table 9-2. Comparison of several amino acids presented in Figure 9-5

Original amino acid	Q	E	R	V	G	W
Possible mutated amino acids	EHKLPR	ADGKQV	CGHIKLMPQSTW	ADEFGILM	ACDERSV W	CGLRS
Number of total mutated amino acids	31	70	101	109	84	4
Mutated amino acid for comparison	H	D	K	I	A	C
Number of mutated amino acid for comparison	9	18	58	76	5	0
Accumulated probability	0.29 (9/31)	0.26 (18/70)	0.57 (58/101)	0.70 (76/109)	0.06 (5/84)	0 (0/4)
Theoretical probability in Table 8-9	0.29	0.29	0.06	0.13	0.17	0.29
Comparison	=	≈	>	>	<	<

Alanine, A; Arginine, R; Aspartic acid, D; Cysteine, C; Glutamic Acid, E; Glutamine, Q; Glycine, G; Histidine, H; Isoleucine, I; Leucine, L; Lysine, K; Methionine, M; Phenylalanine, F; Proline, P; Serine, S; Threonine, T; Tryptophan, W; Valine, V.

229, 239 and 313 in the hemagglutinin of human H3N2 influenza virus (accession number CY016595). Their mutation probability is larger than 0.5 and risks mutation as shown in the bottom panel of Figure 9-6. If these valines mutate, which kind of amino acid will appear?

Two pies in the middle panel of Figure 9-6 display the prediction of would-be-mutated amino acids related to the original amino acid "V". The left pie is constructed according to the theoretical probability in Table 8-9 and the upper panel in Figure 9-5, where "L" has the biggest chance to appear (0.25), followed by amino acids "A" and "G". The right pie is constructed according to the accumulated probability in the lower panel in Figure 9-5, in which vast majority of mutated amino acid is "I". Thus the difference is remarkable between theoretical and accumulated mutating probability.

There are two possibilities for future would-be-mutated amino acids [135-137]. They may continue following the theoretical probability in the left pie, say, the amino acids with bigger probability have larger chance to appear. On the other hand, it is more likely to occur that the amino acids have bigger difference between the theoretical and accumulated probability (the top panel in Figure 9-6), because mutations will follow the theoretical probability in a large-scale and long-term statistics. The mutated amino acids recorded frequently in the past will have less chance to appear in the future. Accordingly, "L" has the biggest chance to appear, followed by "G" but not "A".

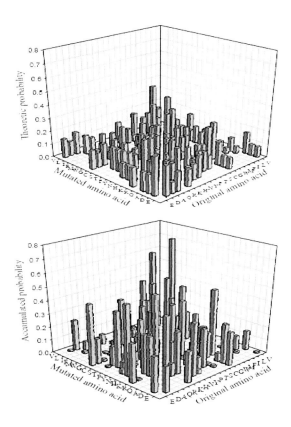

Figure 9-5. Theoretical amino-acid mutating probability (upper panel) and historically accumulated amino-acid mutating probability (lower panel) based on 1346 point mutations from hemagglutinins of EuroAsia human H3N2 influenza viruses.

9.4. Summary of Three Measures

To this moment, we have discussed the rationales of quantifying randomness in protein, we have detailed the calculations with a variety of worked examples, we have examined the behaviors of three measures of randomness from spatial and time angles, and we have shown some applications to research.

Before going ahead, it would be very useful for us to compare these three measures to get several simple taking-home messages rather than endless computation. In Table 9-3, we show the synopsis of three measures of randomness in a protein. We only present the primary measures in this table, as you have already known that there are several derivates of these measures for different purposes, for example, distribution rank versus distribution probability, future mutated amino-acid composition versus future amino-acid composition, etc.

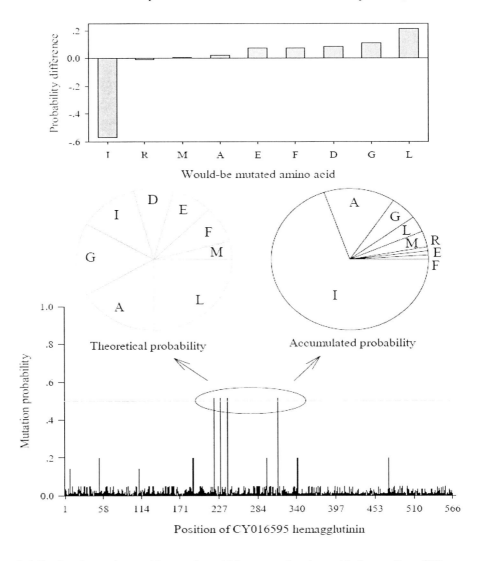

Figure 9-6. Predicted mutation positions and would-be-mutated amino acids from valines "V" at positions 220, 229, 239 and 313 in CY016595 hemagglutinin.

Table 9-3. Synopsis of three computational mutation approaches

Approach	Amino-acid pair predictability	Amino-acid distribution probability	Amino-acid mutating probability
Underlying principle	Permutation	Occupancy of subpopulations and partitions	Translation probability between RNA codons and translated amino acids
General role	Classifying amino-acid pairs in a protein as predictable and unpredictable.	Estimating distribution complexity of amino acids in a protein.	Determining future amino-acid composition in a protein; Predicting would-be mutated amino acids.
Terminology	Pair: an amino-acid pair is composed of two neighboring amino acids. Type: 20 kinds of amino acids construct 400 possible types of amino-acid pairs. Present or absent type: a type of amino-acid pair appears or does not appear in a protein. Predictable/unpredictable type: the presence/absence of a type of amino-acid pair can/cannot be predicted by random principle. Predictable/unpredictable frequency: the appearing number of a type of amino-acid pair can/cannot be predicted by random principle. Predictable/unpredictable portion: the percentage of all predictable/unpredictable types (frequencies) of amino-acid pairs. Difference between actual and predicted frequency: (actual frequency)–(predicted frequency). Type mutation: 1% type mutation is equal to the mutations occurred in 4 types of amino-acid pairs. Frequency mutation: 1% frequency mutation is equal to the mutations occurred in 1% amino-acid pairs in the protein.	Distribution probability: the probability is calculated according to the positions of a kind of amino acids in a protein. Distribution rank: the descending order is sorted from the distribution probability. Distribution rank per amino acid: (distribution rank)/(number of corresponding amino acids). Distribution rank in a protein: (sum of all distribution ranks per amino acid)÷(number of amino-acid kinds in a protein). Predictable portion: the percentage of all amino acids whose actual distribution probabilities equal to the predicted ones. Unpredictable portion: the percentage of all amino acids whose actual distribution probabilities are different from the predicted ones.	Mutating probability: the probability is calculated according to the options from original amino acids to mutated amino acids. Mutating probability including or excluding self-oriented mutation: the mutating probability includes or excludes the options that the mutated amino acids are the same as the original ones. Future composition: the amino-acid composition is calculated according to the mutating probability including self-oriented mutation. Future mutated composition: the amino-acid composition is calculated according to the mutating probability excluding self-oriented mutation. Would-be mutated amino acid: the amino acid will appear after mutation.

Table 9-3. Continued

Approach	Amino-acid pair predictability	Amino-acid distribution probability	Amino-acid mutating probability
Implication	The larger the unpredictable portion is, the less stable the protein is. The larger the unpredictable portion is, the more sensitive to mutation the protein is. The larger the difference between actual and predicted frequency of amino-acid pair is, the stronger the mutation trend is.	The larger the distribution rank is, the less stable the protein is. The larger the distribution rank is, the more probabilistically complicated the protein is. The larger the distribution rank is, the more sensitive to mutation the protein is.	The larger the difference between current and future amino-acid composition is, the stronger the mutation trend is. The larger the difference between theoretical and accumulated mutating probability for a given amino acid is, the larger the chance of appearance of a would-be-mutated amino acid is.

Chapter 10

Prediction of Mutation Position Using Logistic Regression

One of important and exciting applications of computational mutation, as its name suggests, is to predict the mutation. If we could accurately, precisely and reliably predict the mutations in proteins from influenza A virus as well as HIV, we would be in the position to engineer new vaccines against them, and we would be in the position to simulate the evolutionary process, which can be explained by random mechanism.

With respect to the prediction of mutation in proteins, the computational mutation approach could in principle deal with all the aspects of prediction at primary structure, i.e. the prediction of mutation position, the prediction of would-be-mutated amino acids at predicted positions, and the timing of mutation.

The timing of mutation in Section 4.3 is somewhat an empiric approach, because our prediction is mainly based on the pattern comparison, even we have used the fast Fourier transform to find the periodicity of historical mutations. This means that our prediction is based on a dynamic model, because we cannot find out all of the causes accounting for all of the documented mutations.

Here one may wonder why we do not use the randomness as a cause for mutation to build a time-mutation relationship as we have indicated it several times in previous chapters that randomness is a force engineering mutation. Although we have investigated the randomness before and after mutation, this does not warrant us the possibility of building a relationship between time and randomness-engineered-mutation, because the time before and after mutation is related to neither any time unit, such as year, month, day, hour, minute and second, nor calendar time. This is the current difficulty to build the relationship between time and randomness-engineered-mutation.

10.1. Possibility of Building a Prediction Model

On the other hand, we may build a relationship between randomness and mutation as long as it does not deal with time, for example, the prediction of mutation position. The mutation position we mean is the position where an amino acid will mutate, so it is more related to the amino acid at the position rather than the position per se. In fact, the methods developed by us are directly relevant to the randomness in each amino acid in a protein, but not to the exact positions, although the amino-acid distribution probability deals with the positions in a protein.

In view of mutation, each amino acid in a protein has only two possibilities, either mutates or does not mutate. Thus, we can classify the occurrence or non-occurrence of mutation as unity or zero, and build a relationship between randomness and mutation. Naturally everyone hopes this relationship in the equation form as simple as $ax + b = y$, where the randomness is on the left-hand side of equation, the occurrence or non-occurrence of mutation is on the right-hand side of equation, and a and b are the model parameters.

Now let us look at this simplest prediction equation, $ax + b = y$, in order to determine what we need to do with it. In this equation as all the mathematical equations, we generally have the data for the left-hand side of equation. For example, assuming we have $2x + 1 = y$, and we have $x = 1, 2, 3, \ldots$, then we can calculate the results on the right-hand side of equation, $y = 3, 5, 7, \ldots$ This example is the so-called direct problem [142], that is, we know the values of model parameters, $a = 2$ and $b = 1$.

In real-life, we frequently meet the reverse problem, especially in bio-medical sciences. The reverse problem with respect to our simplest equation, $ax + b = y$, is that we know x and y, but we do not know the model parameters, a and b. Of course, with the x and y above, we can easily find the model parameters, a and b. However, the model parameters are very difficult to find in most real-life cases, and the process of finding model parameters is also called as the model identification [142]. Accordingly, let us analyze what problem we are facing and what we have in hands.

No matter whether or not we are dealing with the direct or reverse problem, we must have x, because we need it to input into the equation to calculate y in case of direct problem or we also need x together with y to determine the model parameters, a and b, in case of reverse problem. As we are interested in the prediction of mutation engineered by randomness, x must be one of three quantifications of randomness developed by us. Actually we can easily calculate the quantified randomness if we have a protein.

According to our defined cause-mutation relationship, y is the occurrence and non-occurrence of mutation. We have already known from Chapters 3 and 6 that a mutation is related to two time points, before and after mutation, which in fact are related to two proteins, parent and daughter. Comparing parent and daughter proteins, we can mark each amino acid in parent protein as unity or zero depending on whether or not this amino acid is different from the daughter protein. The comparison of parent with daughter proteins means that these two proteins are historical data. Hence, we can have y by comparing historical data. Meanwhile, we can understand that we will not have y for the proteins that will appear in future because the daughter protein has yet to be mutated from her parent one. So we can conclude that we have historical y but not future y, moreover, we do not have the model parameters, a and b, when building the cause-mutation relationship.

Taking together, we are initially facing the reverse problem to determine the model parameters using historical data, then we are facing the direct problem to use the model with current parent protein to predict the daughter one in future. Technically, x should be related to each amino acid in a parent protein because the comparison between parent and daughter proteins is related to each amino acid in the sequence.

Now let us consider what model we can use for our purpose, certainly $ax + b = y$ is unlikely to be our choice because y does not range between unity and zero.

Fortunately, there are mathematical models whose output ranges between zero and unity, and this sort of modeling belongs to the problem of classification [81, 143, 144]. In general, we can use either logistic regression or neural network to model the output ranging between zero and unity.

10.2. Logistic Regression

Logistic   regression   is   a   statistical   tool   [145],   whose   general   form   is $P(y) = \dfrac{1}{1 + e^{(b_1 x_1 + b_2 x_2 + ... + b_n x_n)}}$. Although the logistic regression equation is far different from our simplest equation in the above section, $ax + b = y$, the meanings are similar, that is, b_i is the model parameters and x_i and $P(y)$ are model input and output, or in terms of regression, x_i and $P(y)$ are independent and dependent.

In our words of prediction, x_i is the causes for mutations, and $P(y)$ is the probability of occurrence or non-occurrence of mutation. The coupling of x_i and $P(y)$ is the job of logistic regression, and the quantitative relationship relies on the model parameters b_i.

As x_i is the causes for mutations, we would expect that the amino-acid pair predictability, amino-acid distribution probability and future amino-acid composition would serve as x_i. Of course, everyone hopes to solve a problem as simple as possible, and now we begin to work on this modeling.

10.2.1. Stepwise Regression of Independents

In regression studies, we have no other choice if we have only one independent. However, we frequently need to find the best combination of independents in order to make the regression equation most powerful, if we have several independents. This is generally done using the stepwise inclusion or exclusion of independents [143].

In influenza database, CY000873 and CY001544 hemagglutinins from North America H3N2 influenza viruses isolated in 2003 can be considered as parent and daughter hemagglutinins. By comparing them, we can mark the amino acid in parent hemagglutinin with zero and unity as non-occurrence and occurrence of mutation.

Table 10-1 details the CY000873 hemagglutinin with our random quantifications and marked mutations. In fact, the comparison between CY000873 and CY001544 hemagglutinins shows six mutations at positions 65, 66, 78, 160, 183 and 363.

Table 10-1. Three quantifications of randomness and occurrence/non-occurrence of mutation in CY000873 hemagglutinin

Position	Amino acid	I	II	III	Mutation
1	M	1	0	1.2243	0
2	K	1	0.1868	0.7099	0
3	T	2	0.9555	1.1644	0
4	I	-2	0.4055	0.7033	0
5	I	-3	0.4055	0.7033	0
6	A	3	1.1479	1.0705	0
7	L	1	0.1699	1.0313	0
8	S	-1	1.7368	1.1012	0
9	Y	1	0.8109	0.9435	0
10	I	0	0.4055	0.7033	0
11	L	2	0.1699	1.0313	0
12	C	3	2.7850	0.8549	0
13	L	4	0.1699	1.0313	0
14	V	3	0.2007	1.4145	0
15	F	1	2.3671	0.8554	0
16	A	3	1.1479	1.0705	0
17	Q	2	1.0704	0.6844	0
18	K	2	0.1868	0.7099	0
19	L	1	0.1699	1.0313	0
20	P	1	1.3863	1.3645	0
21	G	3	1.5759	0.8410	0
22	N	2	0.3417	0.5918	0
23	D	1	1.5994	0.7889	0
24	N	1	0.3417	0.5918	0
25	S	4	1.7368	1.1012	0
26	T	2	0.9555	1.1644	0
27	A	0	1.1479	1.0705	0
28	T	3	0.9555	1.1644	0
29	L	4	0.1699	1.0313	0
30	C	3	2.7850	0.8549	0
31	L	1	0.1699	1.0313	0
32	G	0	1.5759	0.8410	0
33	H	2	0.4700	1.7392	0
34	H	2	0.4700	1.7392	0
35	A	0	1.1479	1.0705	0
36	V	2	0.2007	1.4145	0
37	P	5	1.3863	1.3645	0
38	N	7	0.3417	0.5918	0
39	G	8	1.5759	0.8410	0
40	T	6	0.9555	1.1644	0
41	L	5	0.1699	1.0313	0

Position	Amino acid	I	II	III	Mutation
42	V	4	0.2007	1.4145	0
43	K	2	0.1868	0.7099	0
44	T	2	0.9555	1.1644	0
45	I	1	0.4055	0.7033	0
46	T	0	0.9555	1.1644	0
47	N	1	0.3417	0.5918	0
48	D	2	1.5994	0.7889	0
49	Q	5	1.0704	0.6844	0
50	I	6	0.4055	0.7033	0
51	E	3	5.7918	0.7625	0
52	V	2	0.2007	1.4145	0
53	T	1	0.9555	1.1644	0
54	N	2	0.3417	0.5918	0
55	A	3	1.1479	1.0705	0
56	T	0	0.9555	1.1644	0
57	E	0	5.7918	0.7625	0
58	L	4	0.1699	1.0313	0
59	V	3	0.2007	1.4145	0
60	Q	-1	1.0704	0.6844	0
61	S	2	1.7368	1.1012	0
62	S	6	1.7368	1.1012	0
63	S	6	1.7368	1.1012	0
64	T	6	0.9555	1.1644	0
65	G	4	1.5759	0.8410	1
66	R	3	0.7885	1.5483	1
67	I	3	0.4055	0.7033	0
68	C	3	2.7850	0.8549	0
69	D	3	1.5994	0.7889	0
70	S	1	1.7368	1.1012	0
71	P	2	1.3863	1.3645	0
72	H	5	0.4700	1.7392	0
73	Q	7	1.0704	0.6844	0
74	I	4	0.4055	0.7033	0
75	L	-1	0.1699	1.0313	0
76	D	0	1.5994	0.7889	0
77	G	0	1.5759	0.8410	0
78	E	1	5.7918	0.7625	1
79	N	3	0.3417	0.5918	0
80	C	1	2.7850	0.8549	0
81	T	2	0.9555	1.1644	0
82	L	2	0.1699	1.0313	0
83	I	1	0.4055	0.7033	0

Table 10-1. Continued

Position	Amino acid	I	II	III	Mutation
84	D	1	1.5994	0.7889	0
85	A	3	1.1479	1.0705	0
86	L	5	0.1699	1.0313	0
87	L	2	0.1699	1.0313	0
88	G	0	1.5759	0.8410	0
89	D	0	1.5994	0.7889	0
90	P	2	1.3863	1.3645	0
91	H	3	0.4700	1.7392	0
92	C	3	2.7850	0.8549	0
93	D	3	1.5994	0.7889	0
94	G	3	1.5759	0.8410	0
95	F	4	2.3671	0.8554	0
96	Q	5	1.0704	0.6844	0
97	N	4	0.3417	0.5918	0
98	K	1	0.1868	0.7099	0
99	E	0	5.7918	0.7625	0
100	W	0	0.6286	0.5328	0
101	D	3	1.5994	0.7889	0
102	L	3	0.1699	1.0313	0
103	F	0	2.3671	0.8554	0
104	V	3	0.2007	1.4145	0
105	E	4	5.7918	0.7625	0
106	R	5	0.7885	1.5483	0
107	S	2	1.7368	1.1012	0
108	K	-3	0.1868	0.7099	0
109	A	-1	1.1479	1.0705	0
110	Y	0	0.8109	0.9435	0
111	S	-1	1.7368	1.1012	0
112	N	0	0.3417	0.5918	0
113	C	1	2.7850	0.8549	0
114	Y	1	0.8109	0.9435	0
115	P	1	1.3863	1.3645	0
116	Y	1	0.8109	0.9435	0
117	D	2	1.5994	0.7889	0
118	V	3	0.2007	1.4145	0
119	P	2	1.3863	1.3645	0
120	D	0	1.5994	0.7889	0
121	Y	2	0.8109	0.9435	0
122	V	2	0.2007	1.4145	0
123	S	0	1.7368	1.1012	0
124	L	0	0.1699	1.0313	0
125	R	4	0.7885	1.5483	0

Position	Amino acid	I	II	III	Mutation
126	S	4	1.7368	1.1012	0
127	L	3	0.1699	1.0313	0
128	V	6	0.2007	1.4145	0
129	A	3	1.1479	1.0705	0
130	S	3	1.7368	1.1012	0
131	S	4	1.7368	1.1012	0
132	G	5	1.5759	0.8410	0
133	T	6	0.9555	1.1644	0
134	L	3	0.1699	1.0313	0
135	E	2	5.7918	0.7625	0
136	F	1	2.3671	0.8554	0
137	N	0	0.3417	0.5918	0
138	N	1	0.3417	0.5918	0
139	E	0	5.7918	0.7625	0
140	S	0	1.7368	1.1012	0
141	F	1	2.3671	0.8554	0
142	N	1	0.3417	0.5918	0
143	W	2	0.6286	0.5328	0
144	T	4	0.9555	1.1644	0
145	G	4	1.5759	0.8410	0
146	V	4	0.2007	1.4145	0
147	A	5	1.1479	1.0705	0
148	Q	5	1.0704	0.6844	0
149	N	7	0.3417	0.5918	0
150	G	8	1.5759	0.8410	0
151	T	3	0.9555	1.1644	0
152	S	2	1.7368	1.1012	0
153	S	2	1.7368	1.1012	0
154	A	2	1.1479	1.0705	0
155	C	3	2.7850	0.8549	0
156	K	0	0.1868	0.7099	0
157	R	0	0.7885	1.5483	0
158	R	4	0.7885	1.5483	0
159	S	3	1.7368	1.1012	0
160	N	0	0.3417	0.5918	1
161	K	2	0.1868	0.7099	0
162	S	2	1.7368	1.1012	0
163	F	1	2.3671	0.8554	0
164	F	0	2.3671	0.8554	0
165	S	1	1.7368	1.1012	0
166	R	1	0.7885	1.5483	0
167	L	0	0.1699	1.0313	0

Table 10-1. Continued

Position	Amino acid	I	II	III	Mutation
168	N	1	0.3417	0.5918	0
169	W	1	0.6286	0.5328	0
170	L	0	0.1699	1.0313	0
171	H	3	0.4700	1.7392	0
172	Q	3	1.0704	0.6844	0
173	L	1	0.1699	1.0313	0
174	K	-1	0.1868	0.7099	0
175	N	-1	0.3417	0.5918	0
176	K	2	0.1868	0.7099	0
177	Y	2	0.8109	0.9435	0
178	P	1	1.3863	1.3645	0
179	A	3	1.1479	1.0705	0
180	L	3	0.1699	1.0313	0
181	N	1	0.3417	0.5918	0
182	V	4	0.2007	1.4145	0
183	A	4	1.1479	1.0705	1
184	M	2	0	1.2243	0
185	P	4	1.3863	1.3645	0
186	N	3	0.3417	0.5918	0
187	N	1	0.3417	0.5918	0
188	E	5	5.7918	0.7625	0
189	K	5	0.1868	0.7099	0
190	F	1	2.3671	0.8554	0
191	D	0	1.5994	0.7889	0
192	K	2	0.1868	0.7099	0
193	L	2	0.1699	1.0313	0
194	Y	0	0.8109	0.9435	0
195	I	0	0.4055	0.7033	0
196	W	0	0.6286	0.5328	0
197	G	1	1.5759	0.8410	0
198	V	1	0.2007	1.4145	0
199	H	2	0.4700	1.7392	0
200	H	3	0.4700	1.7392	0
201	P	3	1.3863	1.3645	0
202	G	6	1.5759	0.8410	0
203	T	4	0.9555	1.1644	0
204	D	1	1.5994	0.7889	0
205	S	1	1.7368	1.1012	0
206	D	1	1.5994	0.7889	0
207	Q	5	1.0704	0.6844	0
208	I	5	0.4055	0.7033	0
209	S	1	1.7368	1.1012	0

Position	Amino acid	I	II	III	Mutation
210	L	0	0.1699	1.0313	0
211	Y	0	0.8109	0.9435	0
212	A	2	1.1479	1.0705	0
213	Q	4	1.0704	0.6844	0
214	A	2	1.1479	1.0705	0
215	S	1	1.7368	1.1012	0
216	G	2	1.5759	0.8410	0
217	R	2	0.7885	1.5483	0
218	V	2	0.2007	1.4145	0
219	T	2	0.9555	1.1644	0
220	V	1	0.2007	1.4145	0
221	S	3	1.7368	1.1012	0
222	T	4	0.9555	1.1644	0
223	K	1	0.1868	0.7099	0
224	R	4	0.7885	1.5483	0
225	S	3	1.7368	1.1012	0
226	Q	-1	1.0704	0.6844	0
227	Q	1	1.0704	0.6844	0
228	T	2	0.9555	1.1644	0
229	V	0	0.2007	1.4145	0
230	I	-1	0.4055	0.7033	0
231	P	3	1.3863	1.3645	0
232	N	2	0.3417	0.5918	0
233	I	-1	0.4055	0.7033	0
234	G	0	1.5759	0.8410	0
235	S	1	1.7368	1.1012	0
236	R	1	0.7885	1.5483	0
237	P	2	1.3863	1.3645	0
238	R	3	0.7885	1.5483	0
239	V	1	0.2007	1.4145	0
240	R	1	0.7885	1.5483	0
241	D	1	1.5994	0.7889	0
242	I	1	0.4055	0.7033	0
243	S	4	1.7368	1.1012	0
244	S	4	1.7368	1.1012	0
245	R	3	0.7885	1.5483	0
246	I	3	0.4055	0.7033	0
247	S	1	1.7368	1.1012	0
248	I	0	0.4055	0.7033	0
249	Y	1	0.8109	0.9435	0
250	W	2	0.6286	0.5328	0
251	T	2	0.9555	1.1644	0

Table 10-1. Continued

Position	Amino acid	I	II	III	Mutation
252	I	1	0.4055	0.7033	0
253	V	1	0.2007	1.4145	0
254	K	2	0.1868	0.7099	0
255	P	3	1.3863	1.3645	0
256	G	2	1.5759	0.8410	0
257	D	0	1.5994	0.7889	0
258	I	0	0.4055	0.7033	0
259	L	2	0.1699	1.0313	0
260	L	2	0.1699	1.0313	0
261	I	-1	0.4055	0.7033	0
262	N	0	0.3417	0.5918	0
263	S	4	1.7368	1.1012	0
264	T	6	0.9555	1.1644	0
265	G	4	1.5759	0.8410	0
266	N	-2	0.3417	0.5918	0
267	L	-3	0.1699	1.0313	0
268	I	0	0.4055	0.7033	0
269	A	1	1.1479	1.0705	0
270	P	3	1.3863	1.3645	0
271	R	2	0.7885	1.5483	0
272	G	1	1.5759	0.8410	0
273	Y	1	0.8109	0.9435	0
274	F	1	2.3671	0.8554	0
275	K	1	0.1868	0.7099	0
276	I	1	0.4055	0.7033	0
277	R	5	0.7885	1.5483	0
278	S	5	1.7368	1.1012	0
279	G	1	1.5759	0.8410	0
280	K	1	0.1868	0.7099	0
281	S	4	1.7368	1.1012	0
282	S	3	1.7368	1.1012	0
283	I	1	0.4055	0.7033	0
284	M	3	0	1.2243	0
285	R	6	0.7885	1.5483	0
286	S	4	1.7368	1.1012	0
287	D	0	1.5994	0.7889	0
288	A	1	1.1479	1.0705	0
289	P	0	1.3863	1.3645	0
290	I	-1	0.4055	0.7033	0
291	G	0	1.5759	0.8410	0
292	K	1	0.1868	0.7099	0
293	C	2	2.7850	0.8549	0

Position	Amino acid	I	II	III	Mutation
294	N	2	0.3417	0.5918	0
295	S	3	1.7368	1.1012	0
296	E	2	5.7918	0.7625	0
297	C	3	2.7850	0.8549	0
298	I	3	0.4055	0.7033	0
299	T	0	0.9555	1.1644	0
300	P	3	1.3863	1.3645	0
301	N	7	0.3417	0.5918	0
302	G	4	1.5759	0.8410	0
303	S	0	1.7368	1.1012	0
304	I	0	0.4055	0.7033	0
305	P	3	1.3863	1.3645	0
306	N	4	0.3417	0.5918	0
307	D	1	1.5994	0.7889	0
308	K	1	0.1868	0.7099	0
309	P	1	1.3863	1.3645	0
310	F	2	2.3671	0.8554	0
311	Q	5	1.0704	0.6844	0
312	N	4	0.3417	0.5918	0
313	V	0	0.2007	1.4145	0
314	N	0	0.3417	0.5918	0
315	R	3	0.7885	1.5483	0
316	I	2	0.4055	0.7033	0
317	T	1	0.9555	1.1644	0
318	Y	2	0.8109	0.9435	0
319	G	1	1.5759	0.8410	0
320	A	3	1.1479	1.0705	0
321	C	3	2.7850	0.8549	0
322	P	2	1.3863	1.3645	0
323	R	2	0.7885	1.5483	0
324	Y	2	0.8109	0.9435	0
325	V	3	0.2007	1.4145	0
326	K	2	0.1868	0.7099	0
327	Q	4	1.0704	0.6844	0
328	N	1	0.3417	0.5918	0
329	T	0	0.9555	1.1644	0
330	L	3	0.1699	1.0313	0
331	K	3	0.1868	0.7099	0
332	L	1	0.1699	1.0313	0
333	A	0	1.1479	1.0705	0
334	T	4	0.9555	1.1644	0
335	G	4	1.5759	0.8410	0

Table 10-1. Continued

Position	Amino acid	I	II	III	Mutation
336	M	3	0	1.2243	0
337	R	2	0.7885	1.5483	0
338	N	1	0.3417	0.5918	0
339	V	3	0.2007	1.4145	0
340	P	2	1.3863	1.3645	0
341	E	4	5.7918	0.7625	0
342	K	5	0.1868	0.7099	0
343	Q	2	1.0704	0.6844	0
344	T	0	0.9555	1.1644	0
345	R	-1	0.7885	1.5483	0
346	G	-2	1.5759	0.8410	0
347	I	-3	0.4055	0.7033	0
348	F	-2	2.3671	0.8554	0
349	G	-1	1.5759	0.8410	0
350	A	0	1.1479	1.0705	0
351	I	0	0.4055	0.7033	0
352	A	-1	1.1479	1.0705	0
353	G	1	1.5759	0.8410	0
354	F	2	2.3671	0.8554	0
355	I	2	0.4055	0.7033	0
356	E	4	5.7918	0.7625	0
357	N	6	0.3417	0.5918	0
358	G	5	1.5759	0.8410	0
359	W	1	0.6286	0.5328	0
360	E	1	5.7918	0.7625	0
361	G	2	1.5759	0.8410	0
362	M	2	0	1.2243	0
363	M	2	0	1.2243	1
364	D	2	1.5994	0.7889	0
365	G	2	1.5759	0.8410	0
366	W	2	0.6286	0.5328	0
367	Y	2	0.8109	0.9435	0
368	G	3	1.5759	0.8410	0
369	F	2	2.3671	0.8554	0
370	R	0	0.7885	1.5483	0
371	H	3	0.4700	1.7392	0
372	Q	6	1.0704	0.6844	0
373	N	4	0.3417	0.5918	0
374	S	3	1.7368	1.1012	0
375	E	3	5.7918	0.7625	0
376	G	5	1.5759	0.8410	0
377	T	7	0.9555	1.1644	0

Position	Amino acid	I	II	III	Mutation
378	G	2	1.5759	0.8410	0
379	Q	1	1.0704	0.6844	0
380	A	2	1.1479	1.0705	0
381	A	-1	1.1479	1.0705	0
382	D	2	1.5994	0.7889	0
383	L	4	0.1699	1.0313	0
384	K	2	0.1868	0.7099	0
385	S	4	1.7368	1.1012	0
386	T	3	0.9555	1.1644	0
387	Q	2	1.0704	0.6844	0
388	A	2	1.1479	1.0705	0
389	A	0	1.1479	1.0705	0
390	I	-1	0.4055	0.7033	0
391	N	-1	0.3417	0.5918	0
392	Q	4	1.0704	0.6844	0
393	I	3	0.4055	0.7033	0
394	N	3	0.3417	0.5918	0
395	G	4	1.5759	0.8410	0
396	K	2	0.1868	0.7099	0
397	L	2	0.1699	1.0313	0
398	N	1	0.3417	0.5918	0
399	R	1	0.7885	1.5483	0
400	L	0	0.1699	1.0313	0
401	I	2	0.4055	0.7033	0
402	E	6	5.7918	0.7625	0
403	K	5	0.1868	0.7099	0
404	T	1	0.9555	1.1644	0
405	N	1	0.3417	0.5918	0
406	E	5	5.7918	0.7625	0
407	K	5	0.1868	0.7099	0
408	F	2	2.3671	0.8554	0
409	H	4	0.4700	1.7392	0
410	Q	7	1.0704	0.6844	0
411	I	6	0.4055	0.7033	0
412	E	6	5.7918	0.7625	0
413	K	4	0.1868	0.7099	0
414	E	1	5.7918	0.7625	0
415	F	1	2.3671	0.8554	0
416	S	2	1.7368	1.1012	0
417	E	3	5.7918	0.7625	0
418	V	4	0.2007	1.4145	0
419	E	4	5.7918	0.7625	0

Table 10-1. Continued

Position	Amino acid	I	II	III	Mutation
420	G	2	1.5759	0.8410	0
421	R	3	0.7885	1.5483	0
422	I	1	0.4055	0.7033	0
423	Q	-1	1.0704	0.6844	0
424	D	3	1.5994	0.7889	0
425	L	4	0.1699	1.0313	0
426	E	5	5.7918	0.7625	0
427	K	5	0.1868	0.7099	0
428	Y	3	0.8109	0.9435	0
429	V	5	0.2007	1.4145	0
430	E	3	5.7918	0.7625	0
431	D	-1	1.5994	0.7889	0
432	T	0	0.9555	1.1644	0
433	K	1	0.1868	0.7099	0
434	I	1	0.4055	0.7033	0
435	D	4	1.5994	0.7889	0
436	L	4	0.1699	1.0313	0
437	W	1	0.6286	0.5328	0
438	S	1	1.7368	1.1012	0
439	Y	0	0.8109	0.9435	0
440	N	1	0.3417	0.5918	0
441	A	2	1.1479	1.0705	0
442	E	1	5.7918	0.7625	0
443	L	3	0.1699	1.0313	0
444	L	5	0.1699	1.0313	0
445	V	6	0.2007	1.4145	0
446	A	6	1.1479	1.0705	0
447	L	4	0.1699	1.0313	0
448	E	3	5.7918	0.7625	0
449	N	2	0.3417	0.5918	0
450	Q	0	1.0704	0.6844	0
451	H	0	0.4700	1.7392	0
452	T	1	0.9555	1.1644	0
453	I	2	0.4055	0.7033	0
454	D	4	1.5994	0.7889	0
455	L	1	0.1699	1.0313	0
456	T	-2	0.9555	1.1644	0
457	D	1	1.5994	0.7889	0
458	S	3	1.7368	1.1012	0
459	E	3	5.7918	0.7625	0
460	M	1	0	1.2243	0
461	N	1	0.3417	0.5918	0

Position	Amino acid	I	II	III	Mutation
462	K	3	0.1868	0.7099	0
463	L	2	0.1699	1.0313	0
464	F	0	2.3671	0.8554	0
465	E	1	5.7918	0.7625	0
466	R	0	0.7885	1.5483	0
467	T	0	0.9555	1.1644	0
468	K	0	0.1868	0.7099	0
469	K	0	0.1868	0.7099	0
470	Q	1	1.0704	0.6844	0
471	L	0	0.1699	1.0313	0
472	R	0	0.7885	1.5483	0
473	E	2	5.7918	0.7625	0
474	N	4	0.3417	0.5918	0
475	A	2	1.1479	1.0705	0
476	E	0	5.7918	0.7625	0
477	D	1	1.5994	0.7889	0
478	M	1	0	1.2243	0
479	G	1	1.5759	0.8410	0
480	N	5	0.3417	0.5918	0
481	G	4	1.5759	0.8410	0
482	C	1	2.7850	0.8549	0
483	F	2	2.3671	0.8554	0
484	K	1	0.1868	0.7099	0
485	I	0	0.4055	0.7033	0
486	Y	1	0.8109	0.9435	0
487	H	1	0.4700	1.7392	0
488	K	1	0.1868	0.7099	0
489	C	3	2.7850	0.8549	0
490	D	2	1.5994	0.7889	0
491	N	2	0.3417	0.5918	0
492	A	5	1.1479	1.0705	0
493	C	6	2.7850	0.8549	0
494	I	3	0.4055	0.7033	0
495	G	0	1.5759	0.8410	0
496	S	0	1.7368	1.1012	0
497	I	1	0.4055	0.7033	0
498	R	1	0.7885	1.5483	0
499	N	4	0.3417	0.5918	0
500	G	8	1.5759	0.8410	0
501	T	5	0.9555	1.1644	0
502	Y	2	0.8109	0.9435	0
503	D	1	1.5994	0.7889	0

Table 10-1. Continued

Position	Amino acid	I	II	III	Mutation
504	H	0	0.4700	1.7392	0
505	D	1	1.5994	0.7889	0
506	V	1	0.2007	1.4145	0
507	Y	0	0.8109	0.9435	0
508	R	1	0.7885	1.5483	0
509	D	0	1.5994	0.7889	0
510	E	-2	5.7918	0.7625	0
511	A	2	1.1479	1.0705	0
512	L	3	0.1699	1.0313	0
513	N	0	0.3417	0.5918	0
514	N	1	0.3417	0.5918	0
515	R	1	0.7885	1.5483	0
516	F	2	2.3671	0.8554	0
517	Q	6	1.0704	0.6844	0
518	I	2	0.4055	0.7033	0
519	K	-3	0.1868	0.7099	0
520	G	0	1.5759	0.8410	0
521	V	4	0.2007	1.4145	0
522	E	4	5.7918	0.7625	0
523	L	2	0.1699	1.0313	0
524	K	2	0.1868	0.7099	0
525	S	2	1.7368	1.1012	0
526	G	2	1.5759	0.8410	0
527	Y	1	0.8109	0.9435	0
528	K	-1	0.1868	0.7099	0
529	D	-1	1.5994	0.7889	0
530	W	1	0.6286	0.5328	0
531	I	1	0.4055	0.7033	0
532	L	1	0.1699	1.0313	0
533	W	2	0.6286	0.5328	0
534	I	2	0.4055	0.7033	0
535	S	2	1.7368	1.1012	0
536	F	2	2.3671	0.8554	0
537	A	1	1.1479	1.0705	0
538	I	1	0.4055	0.7033	0
539	S	1	1.7368	1.1012	0
540	C	1	2.7850	0.8549	0
541	F	0	2.3671	0.8554	0
542	L	1	0.1699	1.0313	0
543	L	4	0.1699	1.0313	0
544	C	5	2.7850	0.8549	0
545	I	3	0.4055	0.7033	0

Position	Amino acid	I	II	III	Mutation
546	V	-1	0.2007	1.4145	0
547	L	1	0.1699	1.0313	0
548	L	2	0.1699	1.0313	0
549	G	2	1.5759	0.8410	0
550	F	2	2.3671	0.8554	0
551	I	1	0.4055	0.7033	0
552	M	2	0	1.2243	0
553	W	1	0.6286	0.5328	0
554	A	3	1.1479	1.0705	0
555	C	3	2.7850	0.8549	0
556	Q	0	1.0704	0.6844	0
557	K	-1	0.1868	0.7099	0
558	G	0	1.5759	0.8410	0
559	N	0	0.3417	0.5918	0
560	I	0	0.4055	0.7033	0
561	R	1	0.7885	1.5483	0
562	C	1	2.7850	0.8549	0
563	N	0	0.3417	0.5918	0
564	I	0	0.4055	0.7033	0
565	C	4	2.7850	0.8549	0
566	I	4	0.4055	0.7033	0
567	STOP	1	0	25.7870	0

I, the sum of difference between actual and predicted frequency in two neighbouring amino-acid pairs; II, the natural logarithm of ratio of predicted versus actual amino-acid distribution probability; III, the ratio of future versus current amino-acid composition.

With the data in Table 10-1, we can stepwise try to see whether the logistic regression can capture the cause-mutation relationship hidden between three quantifications of randomness and occurrence/non-occurrence of mutation.

Before trying, we need to give a final look at Table 10-1 to see if anything needs our special attention. Likely, the terminal signal, STOP, is very different from the rest of amino acids in this hemagglutinin, because the ratio of future versus current composition is extremely higher than others. Traditionally, this ratio should be considered as outlier in statistical analysis [146] although we can sometimes see the mutation from STOP to other amino acids in real-life [147]. To avoid the complex, our regression will not include the STOP signal.

Figure 10-1 shows the stepwise regression of three quantifications of randomness with occurrence/non-occurrence of mutation using OS4 software [148]. According to the general consideration in probability, we would expect the occurrence of mutation when the mutation probability is larger than 0.5, which is equivalent to that we toss a coin that is unbalanced so one side of coin would appear more frequently.

In the lower panel of Figure 10-1, we only regress a single quantification of randomness with occurrence/non-occurrence of mutation each time, where we can see that the mutation probability is far smaller than 0.5. This means that a single quantification of randomness is not powerful enough to capture the cause-mutation relationship, thus we need to stepwise increase the quantifications of randomness. On the other hand, it can be seen that the second quantification is the most powerful among three quantifications in the lower panel of Figure 10-1.

In the meddle panel of Figure 10-1, we can see that the mutation probability is larger than that in the lower panel, and the strongest combination is the combination of the first and second quantifications. The results in this panel not only support the stepwise increase in the number of quantifications of randomness, but also imply that our three quantifications of randomness do reflect different aspects of randomness in a protein. Thus we should consider increasing the number of quantifications of randomness again.

In the upper panel of Figure 10-1, we use all three quantifications into the regression. Compared with other two panels, we can see that the mutation probabilities in the upper panel are not large enough to reach the probability of 0.5. Actually there is little difference between middle and upper panels with respect to the mean of all the probabilities. However, we do find that the standard deviation of all the probabilities in the upper panel is larger than that in the middle panel. In other words, the inclusion of three quantifications enlarge the difference between probabilities, this way, we can more easily distinguish the maximum from others.

In fact, the plots in Figure 10-1 include two steps, that is, solving of reverse problem by finding model parameters and solving of direct problem by inputting independents to calculate the dependent with the obtained parameters. However, these two steps are generally done using any statistical software packages for the same dataset.

10.2.2. Derivate Independents

Although we only have defined three quantifications of randomness, we need to consider whether we can include more causes into our cause-mutation relationship at this stage, because our three quantifications of randomness cannot make the mutation probability be larger than 0.5.

The first way is to see if we can define the quantifications, which will change according to the change of three quantifications of randomness. This type of quantifications can be viewed as derivate quantifications, whose role is to increase the sensitivity of logistic regression [143].

In Table 10-1, we notice the first quantification of randomness ranges from –3 to 8. We can count how many –3, –2, –1, until 8, so we have Table 10-2. From this table, we understand that each value of quantification I has different contribution to CY000873 hemagglutinin. On the other hand, we expect that the contribution of quantification I can change when a mutation occurs because of the change of quantification I before and after mutation. In this view, we can assign this contribution to each amino acid in the hemagglutinin sequence as we have done for the quantifications I, II and III, and consider this as a derivate quantification for our logistic regression.

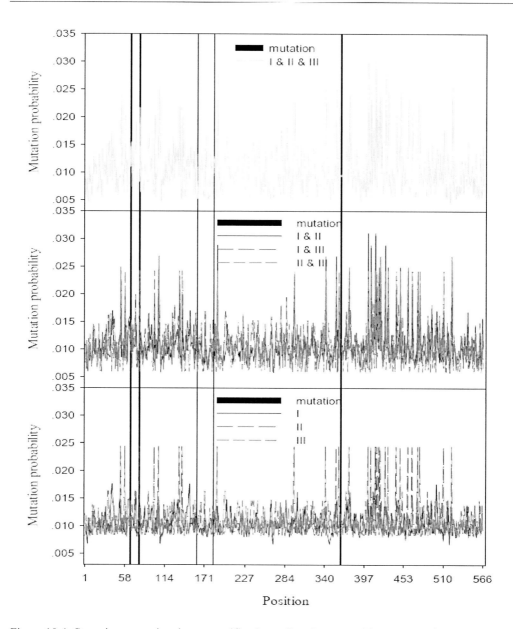

Figure 10-1. Stepwise regressing three quantifications of randomness with occurrence/non-occurrence of mutation. There are six positions marked with mutation but the lines overlap for positions 65 and 66.

By the same token, we can calculate the contribution of each value of quantification II and assign it to CY000873 hemagglutinin sequence because this is also a derivate quantification changes before and after mutation. This way, we now have five quantifications, independents, for our logistic regression.

Another type of derivate quantification, which we can think out, is the first-order interaction between quantifications, which is well documented in statistical literature [143, 144]. In terms of logistic regression, it is the interaction between independents. For us, we can consider the first-order interaction between quantification I and its contribution and between quantification II and its contribution. Furthermore, the first-order interaction is very easy to calculate, quantification I times its contribution and quantification II times its contribution, then we assign the interaction to each amino acid in the hemagglutinin. In this manner, we have seven quantifications of randomness, seven independents, for our logistic regression.

Table 10-2. Contribution of each value of quantification I to CY000873 hemagglutinin

Range of quantification I	Number	Percentage
-3	5	0.88%
-2	6	1.06%
-1	24	4.24%
0	84	14.84%
1	130	22.97%
2	114	20.14%
3	84	14.84%
4	60	10.60%
5	31	5.48%
6	19	3.36%
7	6	1.06%
8	3	0.53%
Total	566	100%

Figure 10-2 shows the regression with three original independents in the lower panel, with three original independents and two contributions in the middle panel and with three original independents, two contributions and two interactions in the upper panel. Although the difference between the lower and middle panels is not so significant, the difference between the largest and smallest mutation probability in the middle panel is bigger than that in the lower panel.

In general, we can see the trend that logistic regression functions better when we include more derivate independents. However, the mutation probability, although increased, is still far lower than the probability of 0.5, which means that we need to further explore the reason to increase the sensitivity of logistic regression.

10.2.3. Sensitivity of Logistic Regression

Another consideration regarding the sensitivity of logistic regression comes from our preliminary study, where we input a fraction of hemagglutinin sequence around 40 amino acids with a mutation into a logistic regression and result in the mutation probability larger than 0.5.

This preliminary study leads us to consider whether the sensitivity of logistic regression depends on the ratio of mutation number versus protein length or the length of protein in question. This notation of course is a scientific hypothesis.

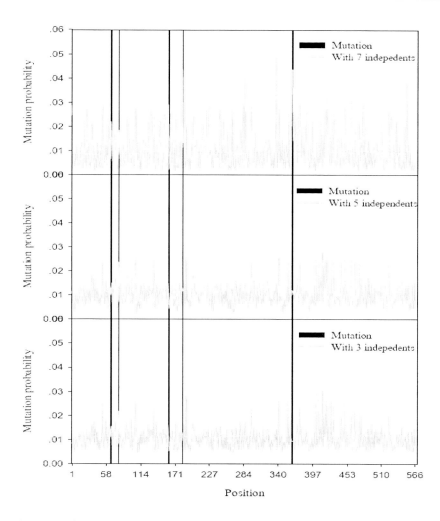

Figure 10-2, Regression using three, five and seven independents with occurrence/non-occurrence of mutation.

Figure 10-3 shows two verifications on this hypothesis using CY000873 hemagglutinin data in Table 10-1. We performed this process by reducing the length of the hemagglutinin step-by-step, and each time we subtracted 10 amino acids from it. From Figure 10-3, we can see that the maximal mutation probability increases as the ratio of mutation number versus protein length increases (the lower panel) or as the length of protein decreases.

The most important implication from Figure 10-3 is that we need to pool more mutations into a single protein in order to increase the power of logistic regression. This in fact is more suitable for the proteins, whose mutations are annotated in the original protein such as p53 protein. However, the proteins from influenza A virus belong to another type of proteins, whose mutations are recorded separately and hardly annotated in a single protein.

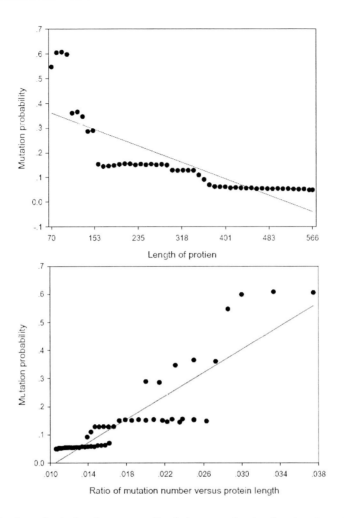

Figure 10-3. Testing hypothesis that the power of logistic regression is related to the ratio of mutation number versus protein length or related to the length of protein.

10.2.4. Pooling of Data

As we have to pool the mutations into a single protein, at best, the original protein, we need to consider pooling the proteins from influenza A virus. Although we cannot identify the original proteins from influenza A virus, we can make the compromise, that is, (i) we align the parent proteins from influenza A virus with the same length, (ii) we calculate the mean of each independent for all parent proteins, (iii) we assign all mutations from all parent proteins into each position, (iv) we thus have a numerical protein composed of the mean of each independent and dependent, and (v) we use this numerical protein into logistic regression to make the prediction.

This compromise is necessary because the logistic regression has the advantage of fast calculation, explicit formulae, and stable model parameters [144]. However, this is not case for other advanced classification tools such as neural network, which generally requires a huge amount of time to try different configurations, with different model parameters, and implicit formulae. Actually, this process is essentially similar to the annotation of all mutations into the original protein such as p53, Cx32 protein.

At the early stage of our studies on the prediction of mutation in proteins from influenza A virus, or the concept-initiated study, we adopted the logistic regression in order to determine whether our cause-mutation relationship works [149-152].

Figure 10-4 illustrates the pooling of 138 parent hemagglutinins of North America H3N2 influenza A viruses isolated from 1968 to 2005 with respect to three quantifications of randomness and 247 mutations. By this format, we can get the model parameters for prediction in order to prove if the cause-mutation relationship works.

Here, the careful reader may ask why we pool the parent hemagglutinins from 1968 to 2005, but not others. This is the issue of sampling strategy, or the prediction based on different datasets, as the database is increasing exponentially. We will address this issue in following section of this Chapter.

10.3. Sampling strategy

10.3.1. Why is the sampling strategy necessary?

With ultra-modern technologies, the database of DNA, RNA and protein is dramatically increasing day-by-day. It is almost impossible to use all the data to make the prediction because this would involve an extreme huge amount of work.

Additionally, the approach of using all the data to make the prediction may not improve the predictability significantly because we can imagine, for example, the model parameters obtained from North America may not be suitable for the prediction of mutation in species lived in EuroAsia.

Here, the problem is from which database to obtain the model parameters is more correctly to predict mutations? Sometimes this issue is called the sampling strategy [153, 154], which is very important for developing timely and economic prediction approaches.

One may wonder if we need to conduct the sampling strategy because the modern phylogenetics can trace the proteins along a branch of evolutionary tree. If this would be the case, we could use the proteins along the branch of evolutionary tree for determining model parameters. However, similar to the fact that we cannot find out all the causes for mutations due to the great changes in environments, we cannot collect all mutant proteins along a branch of evolutionary tree. Therefore, the determined parent-daughter relationship could be the relationship between grandparent and granddaughter. Actually, we could not know how many generations exist between them. In such a case, the sampling strategy appears useful, because it can estimate which dataset is more useful to obtain the relevant model parameters.

Different datasets would affect the mathematical predictions in such a way. The mathematical model we use is a cause-mutation relationship, whose form could conceptually be as simple as $ax + b = y$. As mentioned in Section 10.1, we can calculate x and determine y with historical data, and then obtain the model parameters, a and b. This is a reverse problem, or the model identification. After having a and b from this dataset x and y, we can predict y using the latest protein, x, say, the neuraminidases isolated yesterday. This is a direct problem. Understandably, *a* and *b* based on *x* and *y* of neuraminidases isolated from 2000 may be different from *x* and *y* of neuraminidases from 2007. If this would be the case, we would expect the predictions different. This is the effect of different datasets on the mathematical predictions.

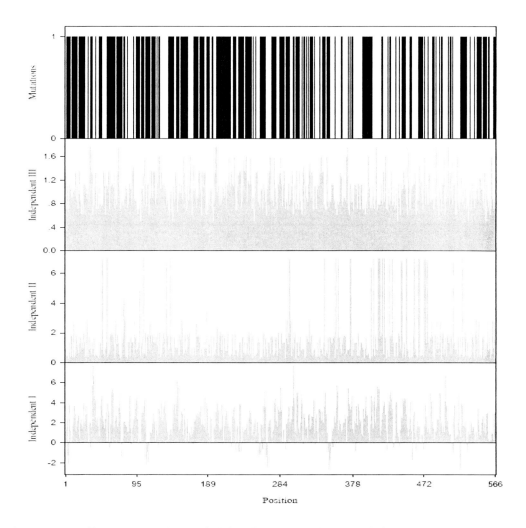

Figure 10-4. Pooling 138 parent hemagglutinins of North America H3N2 influenza viruses from 1968 to 2005.

10.3.2. Stratification of Datasets, Prediction and Comparison

Clearly, the first step for conducting a study on sampling strategy is to stratify database into several subsets based on different considerations [155], for example, based on country where the proteins are found. However, the simplest way is to stratify the database according to the year of isolation because we always make the prediction with the knowledge on the past.

Here we present an example on how to investigate the predictions based on different datasets of H5N1 neuraminidases grouped according to their isolated year. Currently, 428 H5N1 neuraminidases from influenza A viruses in North America are obtained from influenza virus resources [156]. After the comparison, 82 neuraminidases with identical sequences are excluded, while the remaining 346 neuraminidases are considered. Furthermore, the neuraminidases consist of different subgroups regarding their length. It is obvious that the predictions should be made within the neuraminidases with the same or quasi-same length. As vast majority (258) of the neuraminidases possess 449 amino acids, we use them in this study.

First, we group the datasets of 258 neuraminidases according to the year of isolation (from 2000 to 2006). Second, we use each stratified dataset into the logistic regression in order to get the model parameters. Third, we use the obtained parameters from each dataset with quantified randomness to predict the occurrence or non-occurrence of mutation in this dataset and compare with the real occurrence or non-occurrence of mutation. Fourth, using the similar process we predict the occurrence or non-occurrence of mutation in the following-year-dataset and compare with the real occurrence or non-occurrence of mutation. Again we use the simplest equation, $ax + b = y$, to detail this process.

1. We group 258 neuraminidases according to the year of isolation, so we have *x2000* and *y2000*, *x2001* and *y2001*, *x2002* and *y2002*, until *x2000-2006* and *y2000-2006*.
2. We find *a2000* and *b2000*, *a2001* and *b2001*, *a2002* and *b2002*, until *a2000-2006* and *b2000-2006*.
3. We use *a2000* and *b2000* with *x2000* to predict $\hat{y}$ *2000* and compare with *y2000*, until we use *a2000-2006* and *b2000-2006* with *x2000-2006* to predict $\hat{y}$ *2000-2006* and compare with *y2000-2006* (Figure 10-5).
4. We use *a2000* and *b2000* with *x2001* to predict $\hat{y}$ *2001* and compare with *y2001*, until we use *a2000-2005* and *b2000-2005* with *x2006* to predict $\hat{y}$ *2006* and compare with *y2006* (Figure 10-6).

10.3.3. Predictions in Parameter-Generated Dataset

Figure 10-5 illustrates the comparison between actual and predicted mutation based on each dataset with its own model parameters, which can be referred to process (3) in stratification of datasets, prediction and comparison. The figure can be read as follows, the y-axis indicates different datasets from single-year based dataset to multi-year based dataset; and the x-axis indicates the number of mutations of each dataset. As can be seen in Figure 10-5, the actual mutations are quite stable over the years, and solely a single dataset results in no difference between actual and predicted mutation among 28 datasets.

Similarly we have a study, which stratifies 482 hemagglutinins from North America H3N2 influenza A viruses into 96 datasets, and studies the predictions based on these 96 datasets. The results show that the prediction power is limited when the datasets contain the data less than four years, and that the predictions are rapidly deteriorated as the dataset includes the data more than seven years except for the datasets from 1968-1990 to 1968-1996 [153], which gives the further evidence of importance of sampling strategy.

On the other hands, these results clearly do not support to pool the data from too many years together for the prediction. The implication might be that the more detailed collection of recent samples is, the more meaningful for the prediction is.

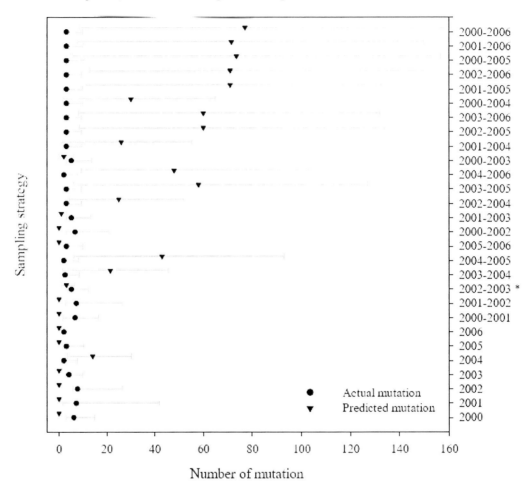

Figure 10-5. Comparison between actual and predicted mutation based on each dataset with its own model parameters. The data are presented as median with an interquartile range. * indicates no statistical difference between actual and predicted mutation.

10.3.4. Predictions in Following-Year Dataset

The importance of prediction is toward future, thus the results in Figure 10-5 can be considered as self-validation. Figure 10-6 displays the comparison between actual and predicted mutation in each dataset with the model parameters obtained from previous dataset, which can be referred to process (4) in stratification of datasets, prediction and comparison. The prediction in Figure 10-6 is nevertheless directed to the future.

The results in Figure 10-6 are somewhat similar to those in Figure 10-5, and only 2 datasets bring about no difference between actual and predicted mutation among 21 datasets.

10.3.5. Patterns of Model Parameters

As we use $P(y) = \dfrac{1}{1 + e^{b_0 + b_1 x_1 + b_2 x_2 + b_3 x_3 + b_4 x_4 + b_5 x_5 + b_6 x_6 + b_7 x_7}}$ for prediction, we should be aware of the patterns of model parameters obtained from different datasets, which are displayed in Figure 10-7. The model parameters, b_i, can be determined analytically using commercially statistical software [148, 157, 158], which is an advantage over the neural network, where the model parameters cannot be determined analytically. As b_i was obtained according to stratification of datasets, the difference between single dataset produced b_i and multi-datasets produced b_i can be seen in Figure 10-7, which can be read as follows. The y-axis indicates different datasets from single-year based dataset to multi-year based dataset, which is as the same as that in Figures 10-5 and 10-6, and the x-axis presents the values of parameters. For example, the parameters based on 2000 dataset are –1.087, 0.236, –0.089, 0.202, –0.003, –1.51, 0.03, and –0.015 for b_0, b_1, b_2, b_3, b_4, b_5, b_6, and b_7, respectively. The parameters based on 2000-2006 datasets are 0.997, –0.209, –0.051, –0.148, –0.038, 0.101, 0.009, and 0.018 for b_0, b_1, b_2, b_3, b_4, b_5, b_6, and b_7, respectively.

As can be seen, the model parameters are unstable when the dataset contains only the data from a single year. Still the model parameters progressively converge as more data are included. These features reveal more remarkable in our study on sampling stratification of H3N2 hemagglutinins [153], and in return support the findings in Figures 10-5 and 10-6.

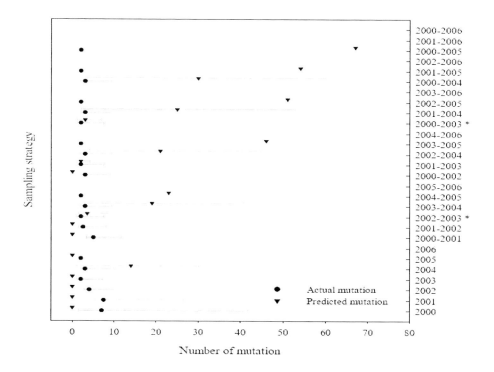

Figure 10-6. Comparison between actual and predicted mutation based on each dataset whose population estimates are derived from the previous year dataset. The data are presented as median with an interquartile range. * indicates no statistical difference between actual and predicted mutation.

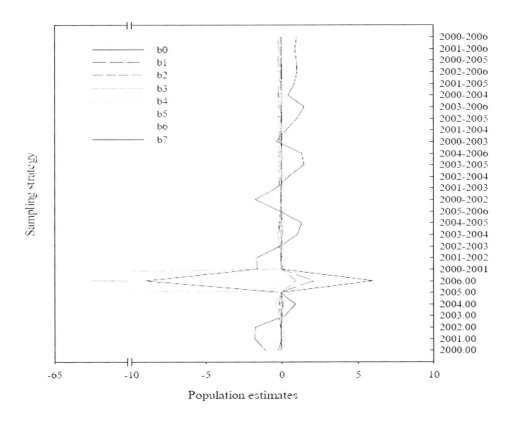

Figure 10-7. Model parameters obtained from different datasets.

Chapter 11

Prediction of Mutation Position Using Neural Network

We have already mentioned that the prediction of mutation positions can be switched to the problem of classification, i.e. to use the continuous numbers to produce the discrete numbers mathematically. Although the logistic regression can model the cause-mutation relationship [149-152], we still need to enhance the predictability not only because the logistic regression is not powerful enough but also, more importantly, because the cause-mutation relationship defined by logistic regression is somewhat simpler.

The formula of logistic regression, $P(y) = \dfrac{1}{1 + e^{(b_1 x_1 + b_2 x_2 + \ldots + b_n x_n)}}$, defines an explicit cause-mutation relationship [144], this is, the causes only affect the output of exponential function, $e^{(b_1 x_1 + b_2 x_2 + \ldots + b_n x_n)}$. However, it is hard to imagine that the nature law, which dictates our daily life, could be found so easily and simply [159]. By contraries, the neural network does not offer any explicit and inflexible cause-mutation relationship, which appears to be more suited to our concept.

From statistical viewpoint, there are two major tools available to solve the problem of classification, one is the logistic regression whose output ranges between zero and unity, and the other is the discriminant analysis whose output can go beyond unity [143]. Unlike the situation in statistics, there are several neural network models, which can do the classification job [81].

At the beginning of applying the neural network to this cause-mutation relationship, we need to consider several aspects on neural network models, datasets and data [159, 160].

11.1. Training Neural Network

11.1.1. Model Selection

Currently, several neural network models are available for classification including the perceptron, backpropagation, probabilistic neutral network, competitive neural network, learning vector quantization networks and self-organizing map [161].

Before conducting the numerical trails, we need to compare these classification models in order to get a concept of which is more suitable for us.

1. The perceptron model is theoretically only suited for the linearly separable problem [161]. This means that the perceptron is somewhat similar to the discriminant analysis in statistics, which is used to classify the linearly separable problem. However, we are more and more aware that the natural phenomena are more likely to be nonlinear, especially, the problem as complicated as mutations, thus we would expect to see the unsatisfied results when using the perceptron model for our cause-mutation relationship although we should confirm this hypothesis with the real-life case.

2. The backpropagation model has been used relatively widely and its mechanism is far more straightforward [162]. It is said that multi-layered networks are capable of performing just about any linear and nonlinear computation, and can approximate any reasonable function arbitrarily well [161]. Therefore, we expected that we could get the satisfied results on nonlinear classification.

3. The probabilistic neutral network [163] once gave us a great expectation because it partially uses the Bayesian approach, which we had elaborated to some degree in Chapter 7. Besides we had other experience on Bayesian application in pharmacokinetics [164-168], thus we would work with this network easily and efficiently. But this network does not require training for model design, which gives us little possibility to improve the model.

4. The competitive neural network and learning vector quantization network belong to the most fascinating topics in neural network [169], however, their classifications depend only on the distance between input vectors. If two input vectors are very similar, the competitive layer probably will put them in the same class. But we have no way to know if our input vectors similar or not.

5. The self-organizing map is more advanced than competitive neural network [161], but we still do not know if it is subject to the requirement for competitive neural network.

Based on the above reasoning, we conducted the preliminary studies to test all these five neural network models. In general, only perceptron and backpropagation can produce meaningful results. In particular, we used 10 hemagglutinins isolated from North America H3N2 influenza viruses for testing different models. The perceptron with a single neuron stopped training after 5 epochs, and reached 0.0131 ± 0.0057 (n = 10) mean squared errors.

By clear contrast, the 3-1 backpropagation reached 0.0081 ± 0.0037 (n = 10) mean squared errors with 250 epochs.

The difficulty in applying the probabilistic neutral network is that this model in principle requires the input of hemagglutinin sequence as many as possible, through which the model parameters are step-by-step trained for all the input hemagglutinin sequences. Certainly, this model is more suitable for a static population of samples, and our results using this model are not satisfied.

Both competitive neural network/learning vector quantization network and self-organizing map face the same problem, namely, the training process is extremely slow because they classify input data, while the length of hemagglutinin is 566, which leads to the very slow convergence.

After these preliminary studies, we decided to use the backpropagation in our studies.

11.1.2. Model Layers and Neurons

After selected the backpropagation model, we need to determine the number of layers for the model as well as the number of neurons for each layer. Because we have three inputs and a target, the easiest way to determine the number of neurons in the first layer of backpropagation model is to choose the number as the same as the number of inputs, while the number of neurons for the output layer should be as the same as the number of target. This way, our neural network already contains two layers with four neurons (3 and 1), which have been tested in the section above.

In fact, the selection of suitable layers and neurons is a process of continuing trials with different layers and neurons. We focus our efforts on the comparison between 3 and 4 layers, because the model parameters, weights and biases, increase rapidly with the increase in layers. For example, the 3-6-1 model (the first, second and third layers contain 3, 6 and 1 neurons, respectively) has 31 weights and biases, the 3-10-1 model has 57 weights and biases, the 3-6-6-1 model has 79 weights and biases, the 3-10-10-1 model has 167 weights and biases. Common knowledge on modeling tells us the less model parameters the better.

With different numbers of layers and neurons, we trained 10 H3N2 hemagglutinins using resilient backpropagation algorithm with 150 epochs, and their results are illustrated in Figure 11-1. In this figure, the x-axis indicates the number of neurons in the second layer of 3-layer model (the filled circles), and the number of neurons in the second and third layers of 4-layer model (the hollow circles).

From the viewpoint of training performance (the upper panel of Figure 11-1), we can find two features: (1) the final mean squared errors reached by 3-layer models are smaller than those reached by the 4-layers; and (2) the model with the best performance has 6 to 8 neurons, which reveal no difference between the 3-layer model and 4-layer one.

From the viewpoint of how many mutations have been captured (the bottom panel of Figure 11-1), the average number of positives is generally more in the 4-layer models than in 3-layer ones. However, this advantage is balanced by the 3-layer model containing 6 neurons. Moreover, the dramatic reduction of weights and biases indeed is attractive, thus we decided to choose the 3-6-1 model.

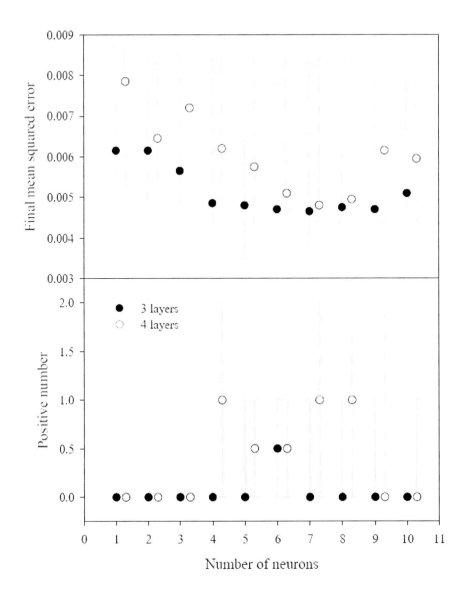

Figure 11-1. Selection of layers and neurons for model. The training is conducted using feedforward backpropagation neural network with 150-epoch resilient backpropagation algorithm. The results are presented as median with interquartile range.

11.1.3. Transfer Functions

Still, we need to define the transfer function for each neuron. There are three transfer functions used for the backpropagation model: (i) the tan-sigmoid function whose output ranges between –1 and 1, (ii) the log-sigmoid function whose output ranges between 0 and 1, (iii) the purelin function whose output ranges from negative infinity to positive infinity. As the occurrence/non-occurrence of mutation ranges between 0 and 1, the transfer function for the third layer should be the log-sigmoid function, and then we can have the output of occurrence and non-occurrence of mutation between 1 and 0.

To evaluate these three transfer functions, we need at least to consider the use of each transfer function in each layer, thus we have the choice for the first two layers because the last one, the output layer, must use the log-sigmoid transfer function, whose output is between 0 and 1. For the first and second layers, each layer has three choices of transfer function, so we have a total of nine combinations.

We evaluate the performance of each transfer function by using the final mean squared error, which is widely used in various fitting settings [143, 144]. Furthermore, we compare the trained output with the real mutation position. In such a case, we have the positive, which is the unity of training output, and the negative, which is the zero of training output. The comparison will tell us which function the best.

Figure 11-2 shows the converging process using different transfer functions in ten hemagglutinins of North America H3N2 influenza viruses. Although the plots in this figure seem to be difficult to find the best, Table 11-1 gives us more details on these comparisons. In general, Figure 11-2 and Table 11-1 favor such a configuration that the first layer can use any transfer function, but the second layer should use the log-sigmoid transfer function that is also used in the third layer.

The results are somewhat different from our previous studies, in which we predicted mutations in proteins of H5N1 viruses, using the configuration that the first and second layers use the tan-sigmoid transfer function and the third layer uses the log-sigmoid transfer function [170, 171]. Thus, the selecting of transfer function also depends on the data to be modeled.

Table 11-1. Comparison of different transfer functions for modeling in 10 hemagglutinins of North America H3N2 influenza viruses

Transfer function		Final mean squared error		P	C	Legends in
First layer	Second layer	Median	Interquartile range			Figure 11-2
tan-sigmoid	tan-sigmoid	0.00537	0.00410-0.00814	6	2	Solid line tt
tan-sigmoid	log-sigmoid	0.00486	0.00430-0.00702	7	3	Dotted line tl
tan-sigmoid	purelin	0.00561	0.00363-0.00805	5	2	Dash line tp
log-sigmoid	tan-sigmoid	0.00549	0.00353-0.00798	2	2	Solid line lt
log-sigmoid	log-sigmoid	0.00482	0.00443-0.00813	7	3	Dotted line ll
log-sigmoid	purelin	0.00478	0.00353-0.00764	5	4	Dash line lp
purelin	tan-sigmoid	0.00585	0.00421-0.00843	4	1	Solid line pt
purelin	log-sigmoid	0.00484	0.00334-0.00761	8	4	Dotted line pl
purelin	purelin	0.00618	0.00525-0.00872	1	1	Dash line pp

The third lay always uses the log-sigmoid transfer function because it ranges from 0 to 1.
P, the positives; C, the correct number.

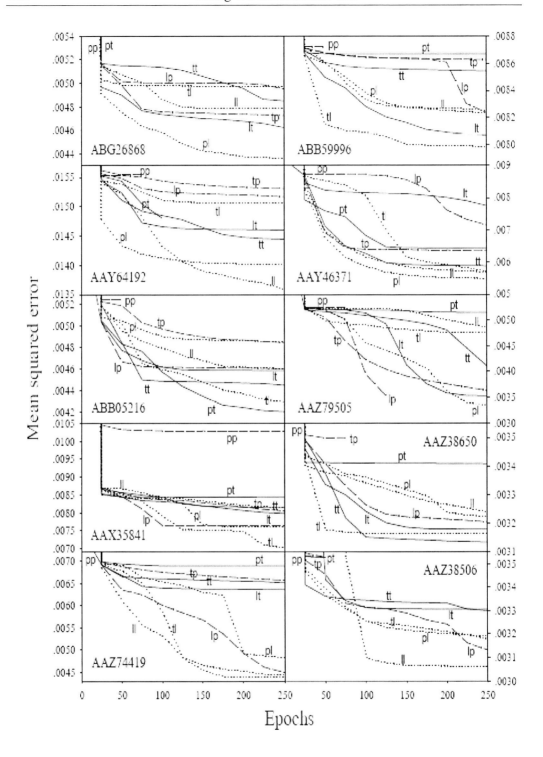

Figure 11-2. Different transfer functions and convergence. The training is conducted using 3-6-1 feedforward backpropagation neural network with resilient backpropagation algorithm. The legends are listed in the last column of Table 11-1.

11.1.4. Training Epochs

For fitting or training, we technically use a mathematical model with different model parameters to approach to experimental data in order that the model output is relatively near to the experimental data [160, 172]. In this sense, the training or fitting is the process that the algorithm changes the model parameters and compares the model output with the experimental data. Clearly, this process may not proceed once, but many times, and each time this process includes the production of new values of model parameters using a particular algorithm, the production of new model output and the comparison between model output and experimental data. In terms of neural network, this process is called epoch [161].

We need to determine the suitable number of epochs for training. If we use too many epochs, although the difference between model output and experimental data may be very small, the model may lack of generalization, say, this model can only be used for this particular dataset, this is the overtraining. By contrast, we also do not hope to have the undertraining, which indicates the big difference between model output and experimental data [161].

Therefore, we should conduct a preliminary study to determine how many epochs are suitable, which means the difference between model output and experimental data stable regardless the number of epochs. Commonly, this difference is measured using the mean squared error because the difference can be either positive or negative. Thus this difference is generally squared although there are some methods using the absolute difference whose speed of convergence is slower than that of squared difference [173].

Figure 11-3 shows the mean squared error versus number of epochs in ten hemagglutinins isolated from North America H3N2 influenza viruses. As seen in this figure, the mean squared error varies at initial point (top of each plot), which is understandable because each hemagglutinin is different one another and its three random measures are different one another, too.

Figure 11-3 can be interpreted in such a way. Each line presents a training process itself with different epochs from 50 to 500. Technically, we train the neural network with different epochs and record the mean squared error with an increment of 25 epochs, i.e. first we use 50 epochs to train the neural network and record the initial, 25-epoch and final mean squared errors, second we use 100 epochs to train the neural network and record the initial, 25-epoch, 50-epoch, 75-epoch and final mean squared errors, until 500 epochs.

From Figure 11-3, we can see that each training generally has a rapid and steep decline in terms of mean squared error, and then the mean squared error decreases slow and finally becomes stable. In general, 250 epochs would be a good choice for training a network because most trainings enter the stable phase before 250 epochs. Moreover, the training in AAZ38650 hemagglutinin has a case that the mean squared error suddenly increases after 300 epochs, which in fact can be seen frequently in various fittings in pharmacokinetics as well as others because the algorithm finds no decrease in mean squared error along the previous converging direction, and then changes its converging direction suddenly [172]. Therefore, the choice of 250 epochs is certainly a good and safe choice.

11.1.5. Training Algorithms

In the above section, we mentioned the algorithm as we were discussing the epochs for training. The algorithm, in plain words, is the mathematical method of how to find the suitable model parameters in a fast and economic fashion [172].

In our studies, we determine the performance of different algorithms after answered questions in above sections such as the number of layers, number of neurons, number of epochs for training, etc. At this stage, we use the feedforward backpropagation neural network to determine the suitable algorithm.

There are in fact many algorithms developed in this field, or there are many variations of the backpropagation algorithm. Any algorithm updates the network weights and biases in the direction in which the performance function decreases most rapidly, that is, the negative direction of gradient. In MatLab, there are two different ways implementing this gradient descent algorithm: incremental mode and batch mode. In the incremental mode, the gradient is computed and the weights are updated after each input is applied to the network. In the batch mode, all of the inputs are applied to the network before the weights are updated [161].

We use the batch mode in our studies [170, 171, 174], because the ratio of mutation number versus protein length is very small, say, we generally have a very few mutations in a hemagglutinin. If we would use the incremental mode, the weights would be largely adjusted in many positions of hemagglutinin, where no mutations are recorded. This would not only affect the efficiency of training in neural network, but also the quantified randomness in the position just before mutation position would weigh much because it would update the model weight lastly.

Furthermore, the training algorithms fall into two main categories. The first category uses heuristic techniques, which are developed from an analysis of the performance of the standard steepest descent algorithm, such as gradient descent, variable learning rate, and resilient backpropagation. The second category uses standard numerical optimization techniques, for example, conjugate gradient, quasi-Newton, and Levenberg-Marquardt [161].

Figure 11-4 displays the training performance using five major algorithms, which are widely used in different fitting and training contexts, and Figure 11-5 shows the training performance using four variations of conjugate gradient algorithms. The training performance is evaluated using 10 hemagglutinins isolated from North America H3N2 influenza viruses.

Due to the difficulty in using nine different lines to present the nine algorithms, we have to present each algorithm one-by-one. In general, each algorithm works quite well, although there are some differences. Thus, we chose the resilient backpropagation, because it is the fastest algorithm on pattern recognition.

11.1.6. Initial Weights and Biases

Once again, for training and fitting we have to choose the initial values for model parameters no matter what model we plan to use. This is because the model does generally

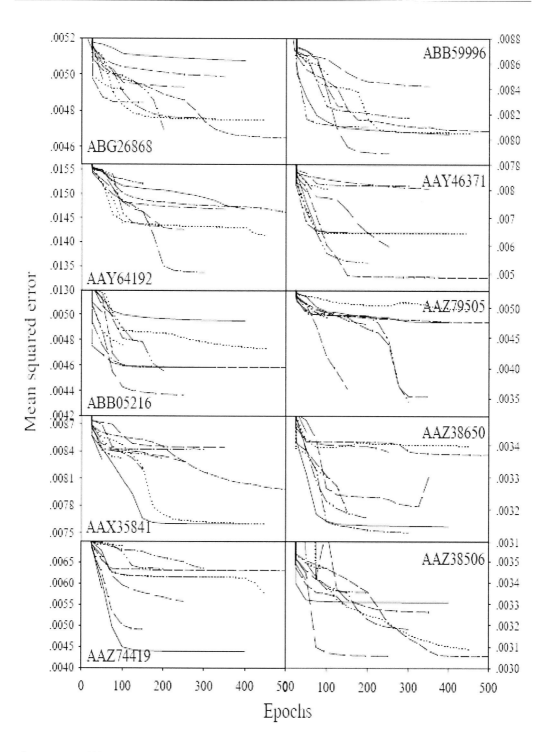

Figure 11-3. Different epochs and convergence. The training is conducted in 10 hemagglutinins of H3N2 influenza viruses, using 3-6-1 feedforward backpropagation neural network with resilient backpropagation algorithm.

not have an explicitly analytical solution when we use the training or fitting to find the model parameters, therefore we need to find a set of initial values for the model parameters. For some model, it is relatively easy to find initial values for model parameters, for example, we can use the graphic method to find the initial values for pharmacokinetic parameters in compartmental model [174, 176], or we can find the initial values from scientific literature on similar problem [176]. However, both are not case for our prediction of mutation position.

Fortunately the neural network program has a function, which produces the initial values for neural network parameters, weights and biases, according to the random mechanism [161]. For us, the problem would be whether the training converges when using the randomly initial values produced by initial function.

Figure 11-6 shows the convergence of mean squared error performance function with different initial weights and biases generated by random initialization function in training 10 H3N2 hemagglutinins. It can be seen that the neural network can converge during its training within 100 epochs although the initial weights and biases were randomly given by the initialization function. Hence, we can safely use the random initialization function to train the neural network to find the model weights and biases.

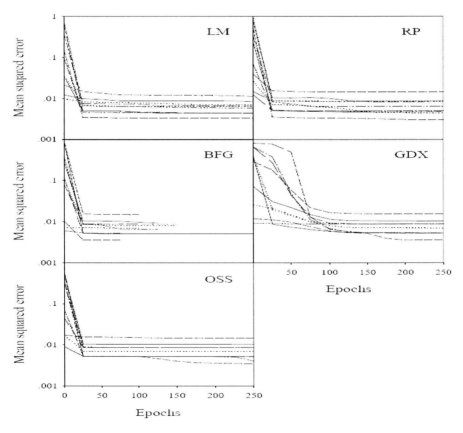

Figure 11-4. Convergence. The training is conducted in 10 H3N2 hemagglutinins by different algorithms in feedforward backpropagation neural network. LM, Levenberg-Marquardt; BFG, BFGS Quasi-Newton; OSS, One-Step Secant; RP, Resilient Backpropagation; GDX, Variable Learning Rate Backpropagation.

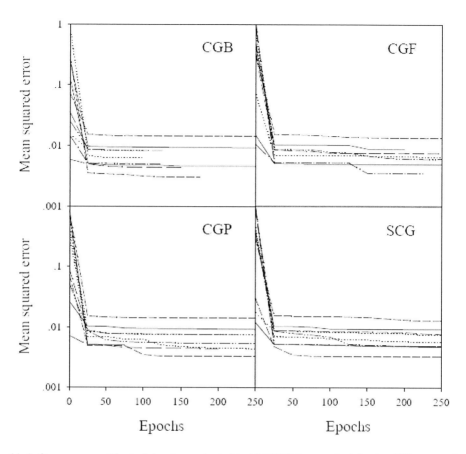

Figure 11-5. Convergence. The training is conducted in 10 H3N2 hemagglutinins by different conjugate gradient algorithms in feedforward backpropagation neural network. CGB, Conjugate Gradient with Powell/Beale Restarts; CGF, Fletcher-Powell Conjugate Gradient; CGP, Polak-Ribiére Conjugate Gradient; SCG, Scaled Conjugate Gradient.

11.1.7. Training, Test and Validation Subsets

The next important issue in our study is whether we need to divide our data into training, test and validation subsets as traditionally done in neural network modeling [161]. At this stage, we do not consider such division necessary. The aim of test and validation subsets is to test whether the model parameters obtained from training subset is valid for test and validation subsets. This division is more suitable for the well-documented field, where there are few problems related to the evaluation of prediction performance.

However, this is not the case for the new developing field of evaluating the prediction of mutation position. We traditionally use the linear regression to make the comparison between actual and predicted value. For instance, in pharmacokinetics we use the measured and predicted blood drug concentrations in x- and y-axes, and then we use the linear regression to see the correlation coefficient [177-179]. However, this evaluation is not possible for the

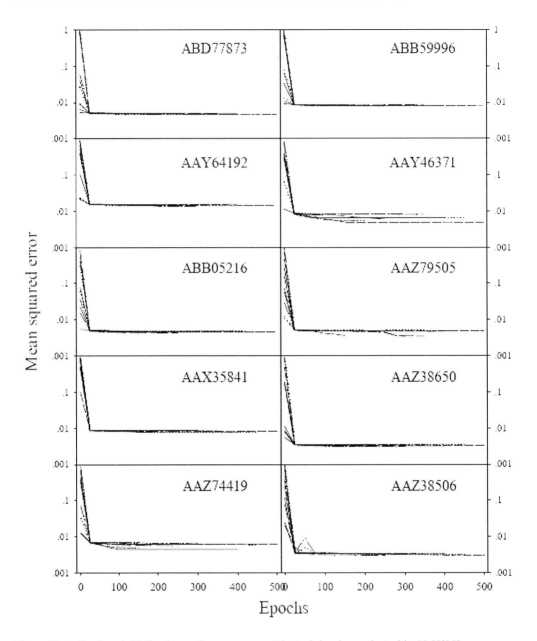

Figure 11-6. Random initialization and convergence. The training is conducted in 10 H3N2 hemagglutinins, using 3-6-1 feedforward backpropagation neural network with resilient backpropagation algorithm.

prediction of mutations, where the situations are more complicated, for example, (i) the actual and predicted mutation positions may not be paired, i.e. there are four actual mutations, but three predicted ones; (ii) the actual and predicted mutation positions are different; and (iii) the actual and predicted mutated amino acids are different.

Therefore, it is too early to divide the dataset into training, test and validation subsets before solving the problems of evaluation of prediction performance.

11.1.8. Preprocessing Data

Finally, the question is whether we need to preprocess data before inputting them into the neural network. In general, there are two methods for preprocessing data in neural network: (i) to scale inputs and targets so that they fall in the range from –1 to 1, and (ii) to normalizes the inputs and targets so that they will have zero mean and unity standard deviation [161].

However, we do not consider that we need to preprocess data because the defaults of preprocessing deal with both inputs and target, and our target is already binary data ranging from zero to unity. Any preprocessing of target will lead to the deformation of target.

In fact, we have already processed input II using the natural logarithm but not processed inputs I and III, because input II is not in the same magnitude regarding inputs I and III, we therefore rescale input II in order to balance all three inputs in the same magnitude. Moreover, the transfer function, tan-sigmoid, in the first and second layers also play the role of preprocessing inputs between –1 and 1. Therefore, we do not need to conduct the preprocessing of our input data.

11.2. Neural Network Prediction

11.2.1. Neural Network Model

Now we are in the position to use the neural network to predict the mutation position after got the answers to the questions related to neural network.

We finally use the feedforward backpropagation neural network as prediction model [161], whose structure is 3-6-1 (Figure 11-7), i.e. the first layer contains three neurons corresponding to three input (or three elements of input in neural network terminology), the second layer contains six neurons, and the last layer contains one neuron corresponding to the target. The transfer functions for three layers are tan-sigmoid, tan-sigmoid and log-sigmoid, respectively. The training algorithm is the resilient backpropagation, which is the fastest algorithm on pattern recognition [161].

11.2.2. Convergence Of Initialization

Although we have showed the convergence in last section, we also can determine this issue in another way, that is, to use the random initialization function to train a single hemagglutinin for 100 times.

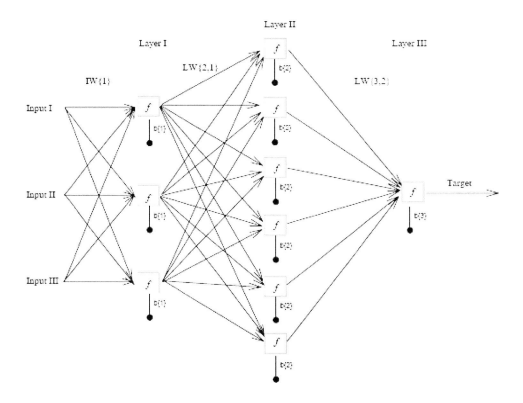

Figure 11-7. The 3-6-1 feedforward backpropagation neural network: each square presents a neuron; IW{1} is the input weights, LW{2,1} is the layer weights to the second layer from the first one, and LW{3,2} is the layer weights to the third layer from the second one; b{1}, b{2} and b{3} are the biases related to each neuron at the first, second, and third layers, respectively; F is the transfer function.

Figure 11-8 shows the convergence of mean squared error performance function with 100 different initial weights and biases generated by random initialization function. We can see that the neural network can converge during its training using ABG88817 hemagglutinin within 250 epochs although the initial weights and biases were randomly given by the initialization function. Hence, we can use the random initialization function to train the neural network to find the suitable weights and biases.

11.2.3 Prediction performance

We have already discussed the difficulty in evaluating prediction performance in relation to common method, linear regression. To go around this difficulty, we use the prediction sensitivity, specificity and total correct rate for the evaluation (Figure 11-9) because we can classify the predicted mutation positions as the positives, false positives, negatives and false negatives when comparing the predicted mutation positions with the actual ones.

The calculations of prediction sensitivity, specificity and total correct rate are according to the published method [158]. From Figure 11-9, we can see that the prediction sensitivity is about 50%, which means that a half of predictions are correct.

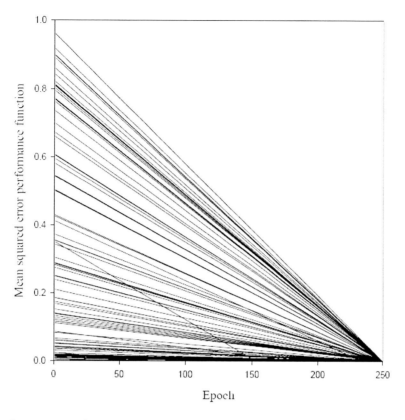

Figure 11-8. Convergence of mean squared error performance function with 100 different initial weights and biases generated by random initialisation function in training ABG88817 hemagglutinin.

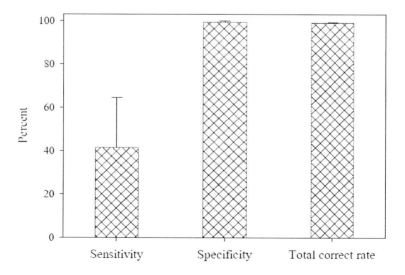

Figure 11-9. Prediction sensitivity, specificity and total correct rate for the self-validation. The data are presented as mean±SD (n=62). The sensitivity is equal to predicted positives/actual mutations (%), the specificity is equal to predicted negatives/actual non-mutations (%), and the total correct rate is equal to (predicted positives + predicted negatives)/length of hemagglutinin (%).

Likely, we can furthermore improve the sensitivity in future using the approaches, which will be discussed in the next chapter. Another way to improve the sensitivity is to find the real parent-daughter relationship, because the current record in conjunction with phylogenetics might yet find the real parent-daughter relationship due to missing samples in history. On the other hand, we can see that the specificity and total correct rate are quite high. These provide the basis for correct prediction of mutations [170, 171, 174].

11.2.4. Prediction of Mutation

As we described earlier that we divide the prediction of mutation into several levels, both the logistic regression and neural network are related to the prediction of mutation position. Thereafter we need to predict the would-be-mutated amino acids at the predicted mutation positions, for which we use the amino-acid mutating probability (Table 8-9).

Figure 11-10 illustrates the two-step frame for the predictions of mutation positions and would-be-mutated amino acids at the predicted positions. In this figure, the solid line in the lower panel is the predicted mutation probability using the neural network for the future mutation positions with respect to CY006076 hemagglutinin and the grey dash line is the cut-off mutation probability of 0.5, that is, the amino acid whose mutation probability is larger than 0.5 risks mutation. In the hemagglutinin, there is only one position whose mutation probability is larger than 0.5, so isoleucine "I" at this position would have a larger chance of occurring of mutation. Meanwhile, the would-be-mutated amino acid at the predicted position can be determined using the amino-acid mutating probability in the upper panel, where we can see the different mutating probabilities related to isoleucine "I".

Our approach is promising because it is based on the internal forces driving mutations, and is different from the current methods, which are more or less based on the phenomenon law, such as searching the similar structures, similar patterns, similar signatures, etc., in various databases. On the one hand, the phenomenon observation is very important [180, 181], by which we can build a dynamic model as the Kelper's laws describe the dynamics of planetary motion. On the other hand, the kinetic deduction is also very important, by which we can build a kinetic model as the Newton's laws describe the kinetics of planetary motion. Moreover, the dynamic model based on phenomenon observation is more suitable to deal with the repeated events, but is less powerful when dealing with the evolutionary process, which generally cannot be reversed [141]. By contrast, the kinetic model can deal with both repeated events and evolutionary process if we can properly define the driving force behind them. Hence, our approach not only has the advantage of quantifying proteins but also has the advantage of kinetic modeling.

Would-be mutated amino acids with probability

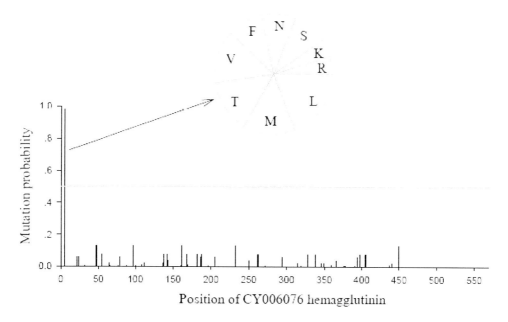

Figure 11-10. Prediction of possible mutation in CY006076 hemagglutinin based on 3-6-1 neural network with IW{1,1}=[0.1999 0.4885 –1.7665; 0.3458 0.4188 1.3805; -1.2788 –6.9395 –0.8043], LW{2,1}=[-82.6000 –24.5000 3.2000; -10.6000 –12.7000 533.7000; -0.3000 1.5000 1.3000; 1.9000 –1.8000 0.6000; -1.3000 –26.0000 1054.2000; -8.3000 –37.5000 1082.3000], LW{3,2}'=[1.5221 1.1972 –1.2557 12.3163 2.0838 0.4014], b{1}=[-0.3712 –2.9052 –2.1586], b{2}=[-5.4124 10.1352 1.0531 –0.0482 25.0149 3.7254], b{3}=[-0.2756], and the amino-acid mutating probability.

Chapter 12

Future Development

During the development of computational mutation approach, we generally follow such a line, i.e. first we propose a new method and formulate its process of calculation, second we prove this new method by observing its behaviors from spatial and time angles, and third we apply this new method to doing research. In particular, Chapters 2, 3 and 4 are dealing with the development, proof and application of amino-acid pair predictability; Chapters 5, 6 and 7 are dealing with the development, proof and application of amino-acid distribution probability; and Chapters 8 and 9 are dealing with the development, proof and application of future composition of amino acids.

The applications in Chapters 4, 7 and 9 are the dynamic analysis because we did not build a cause-response relationship. In general, the dynamic analysis is used to search for the similar patterns along the history line.

Finally we apply three methods together to build a kinetic model in Chapters 10 and 11 to predict the mutation position and predict the would-be-mutated amino acids at the predicted positions in hemagglutinins from influenza A virus.

So far, we have detailed the current state of computational mutation. As a new discipline, we believe that there is a lot of room for its future development, which requires more efforts. Currently, we can see two directions for the future development.

12.1. Distinguishing Arginines, Leucines and Serines

12.1.1. Distinguishing Arginines, Leucines and Serines

At first, let us review once again the unambiguous relationship between RNA codons and translated amino acids shown in Table 8-1, and we present it here again to see what else we can get from it.

Table 12-1. 64 RNA codons and their translated amino acids

UUU UUC	Phenylalanine, F	*UCU* *UCC*	*Serine, S*	UAU UAC	Tyrosine, Y	UGU UGC	Cysteine, C
UUA *UUG*	*Leucine, l*	*UCA* *UCG*		UAA UAG	STOP	UGA UGG	STOP Tryptophan, W
CUU *CUC* *CUA* *CUG*	*Leucine, L*	CCU CCC CCA CCG	Proline, P	CAU CAC	Histidine, H	*CGU* *CGC* *CGA* *CGG*	*Arginine, R*
				CAA CAG	Glutamine, Q		
AUU AUC AUA	Isoleucine, I	ACU ACC ACA ACG	Threonine, T	AAU AAC	Asparagine, N	*AGU* *AGC*	*Serine, s*
AUG	Methionine, M			AAA AAG	Lysine, K	*AGA* *AGG*	*Arginine, r,*
GUU GUC GUA GUG	Valine, V	GCU GCC GCA GCG	Alanine, A	GAU GAC	Aspartic acid, D	GGU GGC GGA GGG	Glycine, G
				GAA GAG	Glutamic acid, E		

One feature that we did not mention in the previous chapters is the amino acids arginine, leucine and serine, of which each corresponds to six RNA codons (marked in bold italic font). What insight can we get from this table? In Table 12-1, these six RNA codons are separated into two groups for each of these three amino acids. We can therefore distinguish them with upper and lower cases of relevant amino acids, that is, we use "R" for the arginine, whose RNA codons are CGU, CGC, CGA and CGG; "r" for the arginine, whose RNA codons are AGA and AGG; "L" for the leucine, whose RNA codons are CUU, CUC, CUA and CUG; "l" for the leucine, whose RNA codons are UUA and UUG; "S" for the serine, whose RNA codons are UCU, UCC, UCA and UCG; "s" for the serine, whose RNA codons are AGU and AGC.

The different groups of RNA codons for the same translated amino acid in Table 12-1 imply that a point mutation in one group of RNA codon may not result in a mutation in the other group.

Let us now look at whether this implication is possible. As we have done in Chapter 8, we list the point mutations at the first, second and third position in RNA codons related to arginine, leucine and serine. Because we are already quite experienced in this type of deduction as Table 8-2, we directly present the results of such deduction in Table 12-2. Similarly, Table 12-2 excludes the self-oriented mutation occurred at RNA codon level, say, we do not consider the case that the RNA codon CGC mutates to CGC.

In Table 12-2, we can see that there are indeed the differences in the probabilities between "R" and "r", between "L" and "l", and between "S" and "s". In plain words, if a mutation is related to "R", we must use the probability related to "R" rather than "r", and similar for the cases of "L" and "l" and of "S" and "s".

Table 12-2. Point mutation at RNA codons and their translated amino acids with distinguishing arginines, leucines and serines

Amino acid	RNA codon	RNA codon position and mutated amino acids			Translation probability
		First position	Second position	Third position	
Arginine, R	CGU	C, G, s	H, L, P	R, R, R	12R+2r+2C+2Q +4G+2H+4L+4P+2s+W+STOP
	CGC	C, G, s	H, L, P	R, R, R	12/36+2/36+2/36+4/36+2/36+4/36+2/36+4/36+2/36+1/36+1/36
	CGA	r, G, STOP	Q, L, P	R, R, R	
	CGG	r, G, W	Q, L, P	R, R, R	
Arginine, r	AGA	R, G, STOP	I, K, T	s, s, r	2R+2r+2G+I+2K+M+4s+2T+W+STOP
	AGG	R, G, W	K, M, T	s, s, r	2/18+2/18+2/18+1/18+2/18+1/18+4/18+2/18+1/18+1/18
Leucine, L	CUU	I, F, V	R, H, P	L, L, L	4R+2Q+2H+3I+2I+12L+M+2F+4P+4V
	CUC	I, F, V	R, H, P	L, L, L	4/36+2/36+2/36+3/36+2/36+12/36+1/36+2/36+4/36+4/36
	CUA	I, I, V	R, Q, P	L, L, L	
	CUG	I, M, V	R, Q, P	L, L, L	
Leucine, l	UUA	I, L, V	S, STOP, STOP	l, F, F	I+2I+2L+M+4F+2S+W+2V+3STOP
	UUG	L, M, V	S, W, STOP	l, F, F	1/18+2/18+2/18+1/18+4/18+2/18+1/18+2/18+3/18
Serine, S	UCU	A, P, T	C, F, Y	S, S, S	4A+2C+2I+2F+4P+12S+4T+W+2Y+3STOP
	UCC	A, P, T	C, F, Y	S, S, S	4/36+2/36+2/36+2/36+4/36+12/36+4/36+1/36+2/36+3/36
	UCA	A, P, T	I, STOP, STOP	S, S, S	
	UCG	A, P, T	I, W, STOP	S, S, S	
Serine, s	AGU	R, C, G	N, I, T	r, r, s	2R+4r+2N+2C+2G+2I+2 s+2T
	AGC	R, C, G	N, I, T	r, r, s	2/18+4/18+2/18+2/18+2/18+2/18+2/18+2/18

Table 12-3. Amino-acid mutating probability with distinguishing arginines, leucines and serines

Original amino acid	Amino-acid mutating probability including and excluding self-oriented mutation at amino-acid level
A, alanine	4/36P+4/36S+4/36T+2/36D+2/36E+4/36G+4/36V+12/36A 4/24P+4/24S+4/24T+2/24D+2/24E+4/24G+4/24V
R, arginine whose RNA codons are CGU, CGC, CGA and CGG	2/36C+2/36r+4/36G+2/36s+1/36STOP+1/36W+2/36H+2/36Q+4/36L+4/36P+12/36R 2/24C+2/24r+4/24G +2/24s +1/24STOP +1/24W +2/24H+2/24Q +4/24L +4/24P
r, arginine whose RNA codons are AGA and AGG	2/18R+2/18G+1/18STOP+1/18W+1/18I+2/18K+1/18M+2/18T+4/18s+2/18r 2/16R+2/16G+1/16STOP+1/16W+1/16I+2/16K+1/16M+2/16T+4/16s
N, asparagine	2/18Y+2/18H+2/18D+2/18I+2/18T+2/18s+2/18N+4/18K 2/16Y+2/16H+2/16D+2/16I+2/16T+2/16s+4/16K
D, aspartic acid	2/18Y+2/18H+2/18N+2/18V+2/18A+2/18G+2/18D+4/18E 2/16Y+2/16H+2/16N+2/16V+2/16A+2/16G+4/16E
C, cysteine	2/18G+2/18R+2/18s+2/18F+2/18S+2/18Y+2/18C+2/18W+2/18STOP 2/16G+2/16R+2/16s+2/16F+2/16S+2/16Y+2/16W+2/16STOP
E, glutamic acid	2/18Q+2/18K+2/18STOP+2/18A+2/18G+2/18V+4/18D+2/18E 2/16Q+2/16K+2/16STOP+2/16A+2/16G+2/16V+4/16D
Q, glutamine	2/18E+2/18K+2/18STOP+2/18R+2/18L+2/18P+2/18Q+4/18H 2/16E+2/16K+2/16STOP+2/16R+2/16L+2/16P+4/16H
G, glycine	6/36R+2/36C+2/36s+1/36STOP+1/36W+4/36A+2/36D+2/36E-4/36V+12/36G 6/24R+2/24C+2/24s+1/24STOP+1/24W+4/24A+2/24D+2/24E-4/24V
H, histidine	2/18N+2/18D+2/18Y+2/18R+2/18L+2/18P+4/18Q+2/18H 2/16N+2/16D+2/16Y+2/16R+2/16L+2/16P+4/16Q
I, isoleucine	3/27L+1/27I+2/27F+3/27V+2/27N+1/27r+2/27s+1/27K+3/27T+6/27I+3/27M 3/21L+1/21I+2/21F+3/21V+2/21N+1/21r+2/21s+1/21K+3/21T+3/21M
L, leucine whose RNA codons are CUU, CUC, CUA and CUG	3/36I+2/36I+2/36F+1/36M+4/36V+4/36R+2/36H+2/36Q+4/36P+12/36L 3/24I+2/24I+2/24F+1/24M+4/24V+4/24R+2/24H+2/24Q+4/24P

Original amino acid	Amino-acid mutating probability including and excluding self-oriented mutation at amino-acid level
l, leucine whose RNA codons are UUA and UUG	1/18I+2/18L+1/18M+2/18V+2/18S+1/18W+3/18STOP+2/18I+4/18F 1/16I+2/16L+1/16M+2/16V+2/16S+1/16W+3/16STOP+4/16F
K, lysine	2/18E+2/18Q+2/18STOP+2/18r+1/18I+1/18M+2/18T+4/18N+2/18K 2/16E+2/16Q+2/16STOP+2/16r+1/16I+1/16M+2/16T+4/16N
M, methionine	1/9I+1/9L+1/9V+1/9r+1/9K+1/9T+3/9I
F, phenylalanine	2/18I+2/18L+2/18V+2/18C+2/18S+2/18Y+4/18I+2/18F 2/16I+2/16L+2/16V+2/16C+2/16S+2/16Y+4/16I
P, proline	4/36A+4/36S+4/36T+4/36R+2/36H+2/36Q+4/36L+12/36P 4/24A+4/24S+4/24T+4/24R+2/24H+2/24Q+4/24L
S, serine whose RNA codons are UCU, UCC, UCA and UCG	4/36A+4/36P+4/36T+2/36C+2/36I+2/36F+3/36STOP+1/36W+2/36Y+12/36S 4/24A+4/24P+4/24T+2/24C+2/24I+2/24F+3/24STOP+1/24W+2/24Y
s, serine whose RNA codons are AGU and AGC	2/18R+2/18C+2/18G+2/18N+2/18I+2/18T+4/18r+2/18s 2/16R+2/16C+2/16G+2/16N+2/16I+2/16T+4/16r
T, threonine	4/36A+4/36P+4/36S+2/36N+2/36r+3/36I+2/36K+2/36s+1/36M+12/36T 4/24A+4/24P+4/24S+2/24N+2/24r+3/24I+2/24K+2/24s+1/24M
W, tryptophan	1/9R+1/9r+1/9G+1/9I+1/9S+2/9STOP+2/9C
Y, tyrosine	2/18N+2/18D+2/18H+2/18C+2/18F+2/18S+2/18Y+4/18STOP 2/16N+2/16D+2/16H+2/16C+2/16F+2/16S+4/16STOP
V, valine	3/36I+4/36L+2/36F+1/36M+4/36A+2/36D+2/36E+4/36G+12/36V 3/24I+2/24I+4/24L+2/24F+1/24M+4/24A+2/24D+2/24E+4/24G

Furthermore we have the amino-acid mutating probability with distinguishing arginines, leucines and serines (Table 12-3). To save the space, we list the amino-acid mutating probability excluding self-oriented mutation at amino-acid level in Table 12-3. The probability in the first line of each cell of the last column is used for the calculation of future amino-acid composition with distinguishing arginines, leucines and serines and that of the second line is used for the future mutated amino acids with distinguishing arginines, leucines and serines. Meanwhile, one may notice that amino acids methionine "M" and tryptophan "W" have only one type of amino-acid mutating probability (Table 12-3), because these two amino acids correspond solely to one RNA codon in Table 12-1.

Now we recap several proteins presented in Table 3-1 in order to compare their composition after distinguishing arginines, leucines and serines (Table 12-4). As seen in Table 12-4, the composition after distinguishing arginines, leucines and serines differs case by case.

Table 12-4. Amino-acid composition with distinguishing arginines, leucines and serines

Protein	AT7B	HBA	LDLR	PH4H	VHL
Accession number	P35670	P01922	P01130	P00439	P40337
Amino acid	Number	Number	Number	Number	Number
R+r	*53*	*3*	*46*	*24*	*20*
R	19	2	21	15	15
r	34	1	25	9	5
L+l	*132*	*18*	*66*	*50*	*20*
L	102	18	56	35	19
l	30	0	10	15	1
S+s	*126*	*11*	*70*	*28*	*11*
S	83	7	41	21	8
s	43	4	29	7	3

R, arginine whose RNA codons are CGU, CGC, CGA and CGG; r, arginine whose RNA codons are AGA and AGG; L, leucine whose RNA codons are CUU, CUC, CUA and CUG; l, leucine whose RNA codons are UUA and UUG; S, serine whose RNA codons are UCU, UCC, UCA and UCG; s, serine whose RNA codons are AGU and AGC.
AT7B, human copper-transporting ATPase 2; HBA, human hemoglobin α-chain; LDLR, human low-density lipoprotein receptor precursor; PH4H, human phenylalanine-4-hydroxylase; VHL, human Von Hippel-Lindau disease tumor suppressor.

12.1.2. Distinguishing in Three Random Measures

Distinguishing arginines, leucines and serines will change the calculations of random measures developed by us in this book.

For the amino-acid pair predictability, distinguishing arginines, leucines and serines suggests that we will have "23" kinds of amino acids rather than 20 ones. Therefore, the possible types of amino-acid pairs will be $23 \times 23 = 529$, which is also our new numerical reference for comparison. For three-amino-acid sequence, the possible types would be $23 \times$

$23 \times 23 = 12167$, although it is less meaningful to count the three-amino-acid sequence in practice.

Similarly, we need to use "23" kinds of amino acids to calculate the amino-acid distribution probability, but this distinguishing warrants an unexpected economic efficiency, as we know in Chapter 5, the distribution patterns increase disproportionally with the increase in amino-acid number. Actually, the distinguishing of arginines, leucines and serines leads to the reduction of distribution patterns needed to calculate.

For the amino-acid mutating probability, we can use Table 12-3 for calculating future amino-acid composition and would-be-mutated amino acids at predicted mutation positions. In order to get the visual sense on the difference between traditional 20 kinds of amino acids and "23" ones with distinguishing arginines, leucines and serines, we can compute the current, future and future mutated amino-acid compositions as done in Figure 8-4. After distinguishing arginines, leucines and serines, we have the amino-acid compositions of human hemoglobin β-chain different from Table 8-7. From Table 12-5, we can see "R" and "l" absent in human hemoglobin β-chain.

Figure 12-1 shows the comparison of current, future and future mutated amino-acid compositions between distinguishing and non-distinguishing of arginines, leucines and serines. When looking at the data in this figure in great details, we found that distinguishing arginines, leucines and serines does have the influence not only on arginines, leucines and serines, but also on other amino acids, such as glutamine "Q", isoleucine "I", phenylalanine "F", proline "P" and STOP signal in this particular protein.

Table 12-5. Amino-acid composition of human hemoglobin β-chain with distinguishing arginines, leucines and serines

Amino acid	Number	Composition (%)
Alanine, A	15	10.2041
Arginine, R	0	0
Arginine, r	3	2.0408
Asparagine, N	6	4.0816
Aspartic acid, D	7	4.7619
Cysteine, C	2	1.3605
Glutamic acid, E	8	5.4422
Glutamine, Q	3	2.0408
Glycine, G	13	8.8435
Histidine, H	9	6.1224
Isoleucine, I	0	0
Leucine, L	18	12.2449
Leucine, l	0	0
Lysine, K	11	7.4830
Methionine, M	2	1.3605
Phenylalanine, F	8	5.4422
Proline, P	7	4.7619
Serine, S	3	2.0408

Table 12-5. Continued

Amino acid	Number	Composition (%)
Serine, s	2	1.3605
Threonine, T	7	4.7619
Tryptophan, W	2	1.3605
Tyrosine, Y	3	2.0408
Valine, V	18	12.2449
Total	147	100

12.1.3. Application of Distinguishing Arginines, Leucines and Serines

In fact, we have already applied this method in two of our recent studies [174, 182]. Both studies use the distinguishing of arginines, leucines and serines for the prediction of mutation positions and would-be-mutated amino acids at the predicted positions, using logistic regression [182] and 3-6-1 feedforward backpropagation neural network [174], respectively.

Figure 12-2 displays the comparison between distinction and non-distinction of arginines, leucines and serines on the prediction of mutation positions using the logistic regression. We evaluate the prediction performance using the percent of captured mutation positions (y-axis) as we said in the previous chapter that the evaluation of prediction performance is still a difficult point in this type of study. Thus, using the percent of captured mutation positions is a good choice for evaluation, as we pooled all the mutations onto a single hemagglutinin sequence.

As clearly showed in Figure 12-2, the prediction based on the distinction of arginines, leucines and serines is better than the prediction based on the non-distinction for both H3N2 and H5N1 hemagglutinins. This is the strong evidence for supporting the distinction of arginines, leucines and serines.

Figure 12-3 illustrates the comparison between the distinction and non-distinction of arginines, leucines and serines on the prediction of mutation positions using the neural network model. By contrast to the prediction made by logistic regression in Figure 12-2, the distinction of arginines, leucines and serines does not show its advantage over the non-distinction. On the one hand, this means that the neural network is indeed very powerful because its training can minimize the difference between distinction and non-distinction. On the other hand, Figures 12-2 and 12-3 can only serve as preliminary studies, and the conclusion cannot be drawn so easily.

Nevertheless distinguishing arginines, leucines and serines is a research direction in future.

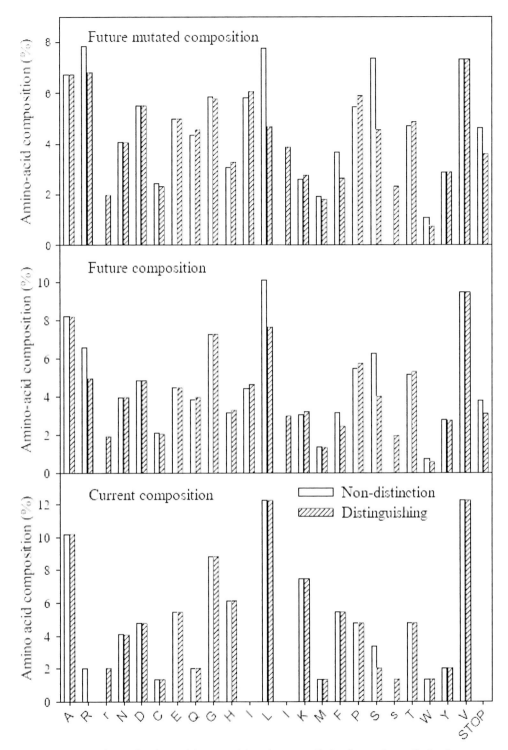

Figure 12-1. Comparison of amino-acid compositions between distinction and non-distinction of arginines, leucines and serines.

12.2. Calculation on RNA Codon Level

As distinguishing arginines, leucines and serines is the direction, we can use it into the relationship between RNA codons and mutated amino acids (Table 12-1), that is, we can compute our three measures with respect to the RNA codon.

Until now, all of our calculations are exclusively focused on the amino-acid level, because at the beginning of our development we did not write the programs that could handle the sequences longer than 1024 symbols, thus we focus on analyzing protein sequences.

An important point is that we do not know from where to start our analysis on DNA sequences, that is, we do not know where a DNA "word" begins, whether there is space between DNA "words", whether there are punctuations, etc.

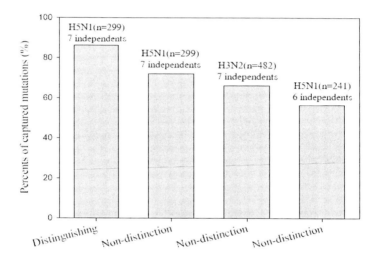

Figure 12-2. Comparison of prediction performance made by logistic regression between distinction and non-distinction of arginines, leucines and serines. The Chi-square test indicates the statistical difference between the predictions made by the distinction and non-distinction.

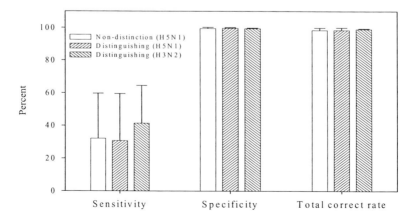

Figure 12-3. Comparison of prediction performance made by neural network between distinction and non-distinction of arginines, leucines and serines.

Still, another point prevented us from doing so was that we could not figure out how to combine DNA or RNA codes in a random format, say, whether we needed to use three codes as a unit for calculating randomness or we needed to use a single code as a unit for the calculation. Also, we did not observe the behaviors of our measures at amino-acid level with respect to time and space.

Very recently, we consider that we can use the RNA codon as a unit for calculating randomness. This means that we use 64 types of RNA codons to calculate the RNA codon pair predictability, RNA codon distribution probability and future composition of RNA codons. In principle, the calculation processes are almost the same as those for the amino-acid pair predictability, amino-acid distribution probability and future amino-acid composition.

For the RNA codon pair predictability, the total number of combination of pairs would increase to 4096 (64×64) possible types of RNA codon pairs from 400 (20×20) possible types of amino-acid pairs. This nevertheless increases the number for classifications in program writing, but more importantly we would expect to see a very few repetitions of RNA codon pairs in a sequence.

Let us have a look at the human hemoglobin α-chain from the view of amino-acid and RNA codon pairs. As can be seen in Table 12-6, the human hemoglobin α-chain has 21 alanines "A", which is the maximal number for amino acids, but has 14 CUGs, which is the maximal number for RNA codons. Consequently, we have such calculations, $21/142 \times 20/141 \times 141 = 2.9577 \approx 3$ for amino-acid pair "AA" and $14/142 \times 13/141 \times 141 = 1.2817 \approx 1$ for RNA codon pairs "CUGCUG", which means that at the best we would expect to one CUGCUG appeared in human hemoglobin α-chain theoretically and randomly. The immediately direct result is that the difference between actual and predicted frequency would be smaller in RNA codon pairs than in amino-acid pairs.

For the distribution probability, the computation would be far less complicated in RNA codons than in amino acids, because the number of each kind of RNA codons is smaller than that of amino acids, thus we do not need to calculate many distribution patterns, which lead to huge amounts of computation.

Table 12-7 lists the comparison in predictable portion (%) between amino-acid and RNA codon levels. As can be seen, (i) the predictable portion far much lower in RNA codon pair predictability than in amino-acid pair predictability, and (ii) the predictable portion is different case by case for RNA codon distribution probability compared with amino-acid distribution probability. Of course, what are shown in Table 12-7 are only the very preliminary results.

For the future RNA codon composition, the situation becomes far more complicated for the preparation for working along this line of thought. More likely, we must have the RNA codon mutating probability, which is similar to Table 8-4, Table 8-9 and Table 12-3, in order to conduct the computation of future RNA codon composition.

The advantage of using RNA codons is that we can take a single code change into consideration, which sometimes does not lead to mutations at amino-acid level. On the other hand, our approach is different from the current approaches on RNA codons [183].

Currently, we have yet to know if we have the need to extend a similar expansion to DNA codes, whether there will be advantage over RNA codons in doing so.

We are sure that we can find more research directions in future with the development of computational mutation.

Table 12-6. Amino-acid and RNA codon compositions of human hemoglobin α-chain

Amino acid	Number	RNA codon	Number	RNA codon	Number	RNA codon	Number	RNA codon	Number
A	21	AAA	1	CAA	0	GAA	0	UAA	1
R	3	AAC	4	CAC	10	GAC	8	UAC	2
N	4	AAG	10	CAG	1	GAG	4	UAG	0
D	8	AAU	0	CAU	0	GAU	0	UAU	1
C	1	ACA	0	CCA	0	GCA	0	UCA	0
E	4	ACC	9	CCC	3	GCC	13	UCC	4
Q	1	ACG	0	CCG	2	GCG	6	UCG	0
G	7	ACU	0	CCU	2	GCU	2	UCU	3
H	10	AGA	0	CGA	0	GGA	0	UGA	0
I	0	AGC	4	CGC	0	GGC	5	UGC	1
L	18	AGG	1	CGG	1	GGG	0	UGG	1
K	11	AGU	0	CGU	1	GGU	2	UGU	0
M	3	AUA	0	CUA	1	GUA	0	UUA	0
F	7	AUC	0	CUC	2	GUC	3	UUC	7
P	7	AUG	3	CUG	14	GUG	9	UUG	0
S	11	AUU	0	CUU	1	GUU	1	UUU	0
T	9								
W	1								
Y	3								
V	13								

For amino acids: A, alanine; R, arginine; N, asparagine; D, aspartic acid; C, cysteine; E, glutamic acid; Q, glutamine; G, glycine; H, histidine; I, isoleucine; L, leucine; K, lysine; M, methionine; F, phenylalanine; P, proline; S, serine; T, threonine; W, tryptophan; Y, tyrosine; V, valine.

Table 12-7. Predictable portions (%) of amino-acid pair and RNA codon pair predictabilities, and of amino-acid and RNA codon distribution probabilities

Protein	AT7B	CA54	GLCM	HBA	LDLR	PH4H	VHL
Accession number	P35670	P29400	P04062	P01922	P01130	P00439	P40337
Amino-acid pair	25.87	26.25	40.23	38.00	30.39	40.17	27.54
RNA codon pair	3.30	2.33	0	0	1.44	0	0
Amino-acid distribution	--	--	--	35.21	--	17.04	11.74
RNA codon distribution	0.75	--	25.37	18.88	1.97	6.18	16.82

AT7B, human copper-transporting ATPase 2; CA54, human collagen α5(IV) chain precursor; FA9, human coagulation factor IX precursor; GLCM, human glucosylceramidase precursor; HBA, human hemoglobin α-chain; LDLR, human low-density lipoprotein receptor precursor; PH4H, human phenylalanine-4-hydroxylase; VHL, human Von Hippel-Lindau disease tumor suppressor. --, the calculations are not conducted.

Appendix

Several Tables on Amino-Acid Distribution Probability and Rank

As there is no simple equation to determine how many distribution patterns for a certain number of amino acids, we list some distribution patterns, their distribution probabilities and ranks in this appendix with reference to the actual distribution of human hemoglobin β-chain. These tables can be used for check the simple distribution patterns for the users who conduct the study along this line of research.

Table 1. Distribution of seven aspartic acids "D" in 7 parts in example of human hemoglobin β-chain

Part 1	Part 2	Part 3	Part 4	Part 5	Part 6	Part 7	Probability	Rank
D	D	D	D	D	D	D	6.1199e-3	8
DD	D	D	D	D	D		0.1285	3
DD	DD	D	D	D			0.3213	1
DD	DD	DD	D				0.1071	4
DDD	*D*	*D*	*D*	*D*			*0.1071*	*4*
DDD	DD	D	D				0.2142	2
DDD	DD	DD					0.0268	6
DDD	DDD	D					0.0178	7
DDDD	D	D	D				0.0357	5
DDDD	DD	D					0.0268	6
DDDD	DDD						1.7850e-3	10
DDDDD	D	D					5.3549e-3	9
DDDDD	DD						1.0710e-3	11
DDDDDD	D						3.5699e-4	12
DDDDDD D							8.4999e-6	13

Bold and italic is the real distribution of human hemoglobin β-chain.

Table 2. Distribution of eight glutamic acids "E" in 8 parts in example of human hemoglobin β-chain

Part 1	Part 2	Part 3	Part 4	Part 5	Part 6	Part 7	Part 8	Probability	Rank
E	E	E	E	E	E	E	E	0.002403	12
	E	E	E	E	E	E	EE	0.0673	5
		E	E	E	E	E	EEE	0.0673	5
			E	E	E	E	EEEE	0.0280	7
				E	E	E	EEEEE	5.6076e-3	9
					E	E	EEEEEE	5.6076e-4	13
						E	EEEEEEE	2.6703e-5	17
							EEEEEEEE	4.7684e-7	18
		E	*E*	*E*	*E*	*EE*	*EE*	*0.2523*	*1*
			E	E	E	EE	EEE	0.2243	2
				E	E	EE	EEEE	0.0421	6
					E	EE	EEEEE	3.3646e-3	11
						EE	EEEEEE	9.3460e-5	16
			E	E	EE	EE	EE	0.1682	3
				E	EE	EE	EEE	0.0841	4
					EE	EE	EEEE	4.2057e-3	10
				EE	EE	EE	EE	0.0105	8
				E	E	EEE	EEE	0.0280	7
					EE	EEE	EEE	5.6076e-3	9
					E	EEE	EEEE	5.6076e-3	9
						EEEE	EEEE	1.1683e-4	15
						EEE	EEEEE	1.8692e-4	14

Bold and italic is the real distribution of human hemoglobin β-chain.

Table 3. Distribution of nine histidines "H" in 9 parts in example of human hemoglobin β-chain

I	II	III	IV	V	VI	VII	VIII	IX	Probability	Rank
1	1	1	1	1	1	1	1	1	9.3666e-4	15
	1	1	1	1	1	1	1	2	0.0337	7
		1	1	1	1	1	1	3	0.0393	5
			1	1	1	1	1	4	0.0197	9
				1	1	1	1	5	4.9174e-3	13
					1	1	1	6	6.5566e-4	16
						1	1	7	4.6833e-5	21
							1	8	1.6726e-6	25
								9	2.3231e-8	26
		1	1	1	1	1	2	2	0.1770	2
			1	1	1	1	2	3	0.1967	1
			1	1	1	1	2	4	0.0492	4
				1	1	1	2	5	5.9009e-3	12
					1	1	2	6	3.2783e-4	20
						1	2	7	6.6904e-6	24
			1	*1*	*1*	*2*	*2*	*2*	*0.1967*	*1*
				1	1	2	2	3	0.1475	3
					1	2	2	4	0.0148	10
						2	2	5	4.9174e-4	17
				1	2	2	2	2	0.0369	6
					2	2	2	3	9.8349e-3	11
				1	1	1	3	3	0.0328	8
					1	2	3	3	0.0197	9
						3	3	3	3.6426e-4	19
					1	1	3	4	9.8349e-3	11
						2	3	4	1.6391e-3	14
						1	4	4	4.0979e-4	18
						1	3	5	6.5566e-4	16
							4	5	2.3416e-5	22
							3	6	1.5611e-5	23

Bold and italic is the real distribution of human hemoglobin β-chain.

Table 4. Distribution of ten amino acids in 10 parts

I	II	III	IV	V	VI	VII	VIII	IX	X	Probability	Rank
1	1	1	1	1	1	1	1	1	1	3.6288e-4	23
2	1	1	1	1	1	1	1	1		0.0163	10
2	2	1	1	1	1	1	1			0.1143	3
2	2	2	1	1	1	1				0.1905	1
2	2	2	2	1	1					0.0714	4
2	2	2	2	2						2.8577e-3	16
3	1	1	1	1	1	1	1			0.0218	9
3	2	1	1	1	1	1				0.1524	2
3	2	2	1	1	1					0.1905	1
3	2	2	2	1						0.0381	6
3	3	1	1	1	1					0.0318	7
3	3	2	1	1						0.0381	6
3	3	2	2							3.1752e-3	15
3	3	3	1							1.4112e-3	19
4	1	1	1	1	1	1				0.0127	11
4	2	1	1	1	1					0.0476	5
4	2	2	1	1						0.0286	8
4	2	2	2							1.5876e-3	18
4	3	1	1	1						0.0127	11
4	3	2	1							6.3504e-3	13
4	3	3								1.5120e-4	25
4	4	1	1							7.9380e-4	21
4	4	2								1.1340e-4	26
5	1	1	1	1	1					3.8102e-3	14
5	2	1	1	1						7.6205e-3	12
5	2	2	1							1.9051e-3	17
5	3	1	1							1.2701e-3	20
5	3	2								1.8144e-4	24
5	4	1								9.0720e-5	27
5	5									1.1340e-6	33
6	1	1	1	1						6.3504e-4	22
6	2	1	1							6.3504e-4	22
6	2	2								4.5360e-5	29
6	3	1								6.0480e-5	28
6	4									1.8900e-6	32
7	1	1	1							6.0480e-5	28
7	2	1								2.5920e-5	30
7	3									1.0800e-6	34
8	1	1								3.2400e-6	31
8	2									4.0500e-7	35
9	1									9.0000e-8	36
10										1.0000e-9	37

References

The following references by no means are the classical references because the advance of science would make these references out-of-date soon. They are those we read during the development of computational mutation and applying our approaches to various problems.

[1]   Impact Factor List 1999. http://www.georgikon.hu/phd/if99.html.

[2]   Top 200 Scientific Journals 1997 and 1998. http://www.phy.hr/~bp/impact.html.

[3]   Szirtes T. *Applied dimensional analysis and modeling*. 1st edition, New York: McGraw-Hill Professional; 1997.

[4]   Gordon G. *System simulation*. 2nd edition. New Jersey: Prentice-Hall; 1978; pp 144-196.

[5]   Sorimachi K, Okayasu T, Ebara Y, Nakagawa T. Mathematical proof of genomic amino acid composition homogeneity based on putative small units. *Dokkyo. J. Med. Sci.* 2005; 32: 99-100.

[6]   Sokal RR, Rohlfe FJ. *Biometry: the principles and practice of statistics in biological research*. 2nd edition. New York: W. H. Freeman; 1981.

[7]   Daniel WW. *Biostatistics: a foundation for analysis in the health sciences*. New York: Wiley; 1991.

[8]   Macheras P., Iliadis A. *Modeling in biopharmaceutics, pharmacokinetics and pharmacodynamics homogeneous and heterogeneous approaches*. Series: *Interdisciplinary applied mathematics*. Vol 30, New York: Springer; 2006.

[9]   Gibaldi M, Perrier D. *Pharmacokinetics*. New York; Marcel Dekker 1982.

[10]   Bairoch A, Apweiler R. The SWISS-PROT protein sequence data bank and its supplement TrEMBL in 2000. *Nucleic. Acids. Res.* 2000; 28: 45-48.

[11]   Chou KC. Structural bioinformatics and its impact to biomedical science. *Curr. Med. Chem.* 2004; 11: 2105-2134.

[12]   Dobson PD, Cai YD, Stapley BJ, Doig AJ. Prediction of protein function in the absence of significant sequence similarity. *Curr. Med. Chem.* 2004; 11: 2135-2142.

[13]   Liu W-M. High density dna microarrays: algorithms and biomedical applications. *Curr. Med. Chem.* 2004; 11: 2143-2152.

[14]   Anfinsen CB, Scheraga HA. Experimental and theoretical aspects of protein folding. *Adv. Protein. Chem.* 1975; 29: 205-300.

[15] Chou KC, Maggiora GM. Domain structural class prediction. *Protein. Eng.* 1998; 11: 523-538.

[16] A short history of geometry. http://www.geometryalgorithms.com/history.htm.

[17] Everitt BS. *Chance rules: an informal guide to probability, risk, and statistics.* New York: Springer; 1999.

[18] Encarta world english dictionary. http://encarta.msn.com/dictionary_/random.html.

[19] http://mw1.merriam-webster.com/dictionary.

[20] Schneider TD. Evolution of biological information. *Nucleic. Acids. Res.* 2000; 28: 2794-2799.

[21] van Tilborg HCA. *An introduction to cryptology.* Boston: Kluwer Academic Publishers; 1989.

[22] Wu G. The first and second order Markov chain analysis on amino acids sequence of human haemoglobin α-chain and its three variants with low O_2 affinity. *Comp. Haematol. Int.* 1999; 9: 148-151.

[23] Wu G, Yan SM. Analysis of distributions of amino acids and amino acid pairs in human tumor necrosis factor precursor and its eight mutations according to random mechanism. *J. Mol. Model.* 2001; 7: 318-323.

[24] Wu G. Frequency and Markov chain analysis of amino-acid sequences of mouse p53. *Hum. Exp. Toxicol.* 2000; 19: 535-539.

[25] Wu G, Yan SM. Prediction of two- and three-amino acid sequence of human acute myeloid leukemia 1 protein from its amino acid composition. *Comp. Haematol. Int.* 2000; 10: 85-89.

[26] Wu G, Yan SM. Prediction of presence and absence of two- and three-amino-acid sequence of human tyrosinase from their amino acid composition and related changes in human tyrosinase variant causing oculocutaneous albinism. *Pediatr. Relat. Top.* 2001; 40: 153-166.

[27] Wu G. The first, second, third and fourth order Markov chain analysis on amino acids sequence of human dopamine β-hydroxylase. *Mol. Psychiatry.* 2000; 5: 448-451.

[28] PROSITE: a dictionary of protein sites and patterns user manual, http://www.expasy.ch/prosite/.

[29] Wu G, Yan S. Determination of amino acid pairs sensitive to variants in human copper-transporting ATPase 2. *Biochem. Biophys. Res. Commun.* 2004; 319: 27-31.

[30] Wu G, Yan S. Analysis of amino acid pairs sensitive to variants in human collagen α5(IV) chain precursor by means of a random approach. *Peptides.* 2003; 24: 347-352.

[31] Wu G, Yan S. Determination of amino acid pairs sensitive to variants in human coagulation factor IX precursor by means of a random approach. *J. Biomed. Sci.* 2003; 10: 451-454.

[32] Wu G, Yan S. Determination of amino acid pairs sensitive to variants in human β-glucocerebrosidase by means of a random approach. *Protein. Eng.* 2003; 16: 195-199. http://peds.oxfordjournals.org/cgi/reprint/16/3/195.

[33] Wu G, Yan SM. Determination of amino acid pairs in human haemoglobulin α-chain sensitive to variants by means of a random approach. *Comp. Clin. Pathol.* 2003; 12: 21-25.

[34] Wu G, Yan S. Determination of amino acid pairs sensitive to variants in human low-density lipoprotein receptor precursor by means of a random approach. *J. Biochem. Mol. Biol. Biophys.* 2002; 6: 401-406.

[35] Wu G, Yan SM. Estimation of amino acid pairs sensitive to variants in human phenylalanine hydroxylase protein by means of a random approach. *Peptides.* 2002; 23: 2085-2090.

[36] Wu G, Yan S. Determination of amino acid pairs in Von Hippel-Lindau disease tumour suppressor (G7 protein) sensitive to variants by means of a random approach. *J. Appl. Res.* 2003; 3: 512-520. http://www.jrnlappliedresearch.com/articles/Vol3Iss4/Wu.pdf.

[37] Wu G, Yan S. Reasoning of spike glycoproteins being more vulnerable to mutations among 158 coronavirus proteins from different species. *J. Mol. Model.* 2005; 11: 8-16.

[38] Wu G, Yan S. Potential targets for anti-SARS drugs in the structural proteins from SARS related coronavirus. *Peptides.* 2004; 25: 901-908.

[39] Wu G, Yan S. Fate of influenza A virus proteins. *Protein. Pept. Lett.* 2006; 13: 377-384.

[40] Wu G, Yan SM. Prediction of distributions of amino acids and amino acid pairs in human haemoglobin α-chain and its seven variants causing α-thalassemia from their occurrences according to the random mechanism. *Comp. Haematol. Int.* 2000; 10: 80-84.

[41] Yan SM, Wu G. Determination of amino acid pairs in human p53 protein sensitive to mutations/variants by means of a random approach. *Chimia.* 2002; 56: 350.

[42] Wu G, Yan S. Fate of 130 hemagglutinins from different influenza A viruses. *Biochem. Biophys. Res. Commun.* 2004: 317; 917-924.

[43] Wu G, Yan S. Determination of amino acid pairs sensitive to variants in human Bruton's tyrosine kinase by means of a random approach. *Mol. Simul.* 2003; 29: 249-254.

[44] Wu G, Yan S. Mutation features of 215 polymerase proteins from different influenza A viruses. *Med. Sci. Monit.* 2005; 11: BR367-BR372. http://www.medscimonit.com/fulltxt.php?ICID=430293.

[45] Wu G, Yan S. Prediction of amino acid pairs sensitive to mutations in the spike protein from SARS related coronavirus. *Peptides.* 2003; 24: 1837-1845. http://www.sibs.ac.cn/sars/file/wenxian/05-12-19.pdf.

[46] Wu G, Yan S. Susceptible amino acid pairs in variants of human collagen α-1(III) chain precursor. *EXCLI J.* 2004; 3: 20-28 http://www.excli.de/vol3/Wu04-04proofrev.pdf.

[47] Wu G, Yan S. Amino acid pairs sensitive to variants in human collagen α 1(I) chain precursor. *EXCLI J.* 2004; 3: 10-19 http://www.excli.de/vol3/Guang03-04.pdf.

[48] Baigent SJ, McCauley JW. Influenza type A in humans, mammals and birds: determinants of virus virulence, host-range and interspecies transmission. *Bioessays.* 2003; 25: 657-671.

[49] Tognotti E. Scientific triumphalism and learning from facts: bacteriology and the "Spanish flu" challenge of 1918. *Soc. Hist. Med.* 2003; 16: 97-110.

[50] Reid AH, Taubenberger JK. The origin of the 1918 pandemic influenza virus: a continuing enigma. *J. Gen. Virol.* 2003; 84: 2285-2292.

[51]   Schafer JR, Kawaoka Y, Bean WJ, Suss J, Senne D, Webster RG. Origin of the pandemic 1957 H2 influenza A virus and the persistence of its possible progenitors in the avian reservoir. *Virology*. 1993; 194: 781-788.

[52]   Bean WJ, Schell M, Katz J, Kawaoka Y, Naeve C, Gorman O, Webster RG. Evolution of the H3 influenza virus hemagglutinin from human and nonhuman hosts. *J. Virol.* 1992; 66: 1129-1138.

[53]   Lin YP, Shaw M, Gregory V, Cameron K, Lim W, Klimov A, Subbarao K, Guan Y, Krauss S, Shortridge K, Webster R, Cox N, Hay A. Avian-to-human transmission of H9N2 subtype influenza A viruses: relationship between H9N2 and H5N1 human isolates. *Proc. Natl. Acad. Sci. U. S. A.* 2000; 97: 9654-9658.

[54]   Katz JM. The impact of avian influenza viruses on public health. *Avian. Dis.* 2003; 47: 914-920.

[55]   Perdue ML, Swayne DE. Public health risk from avian influenza viruses. *Avian. Dis.* 2005; 49: 317-327.

[56]   de la Barrera CA, Reyes-Teran G. Influenza: forecast for a pandemic. *Arch. Med. Res.* 2005; 36: 628-636.

[57]   Hilleman MR. Realities and enigmas of human viral influenza: pathogenesis, epidemiology and control. *Vaccine*. 2002; 20: 3068-3087.

[58]   Wood GW, Banks J, McCauley JW, Alexander DJ. Deduced amino acid sequences of the haemagglutinin of H5N1 avian influenza virus isolates from an outbreak in turkeys in Norfolk, England. *Arch. Virol.* 1994; 134: 185-194.

[59]   Guan Y, Peiris JS, Lipatov AS, Ellis TM, Dyrting KC, Krauss S, Zhang LJ, Webster RG, Shortridge KF. Emergence of multiple genotypes of H5N1 avian influenza viruses in Hong Kong SAR. *Proc. Natl. Acad. Sci. U. S. A.* 2002; 99: 8950-8955.

[60]   Lipatov AS, Krauss S, Guan Y, Peiris M, Rehg JE, Perez DR, Webster RG. Neurovirulence in mice of H5N1 influenza virus genotypes isolated from Hong Kong poultry in 2001. *J. Virol.* 2003; 77: 3816-3823.

[61]   Li KS, Guan Y, Wang J, Smith GJ, Xu KM, Duan L, Rahardjo AP, Puthavathana P, Buranathai C, Nguyen TD, Estoepangestie AT, Chaisingh A, Auewarakul P, Long HT, Hanh NT, Webby RJ, Poon LL, Chen H, Shortridge KF, Yuen KY, Webster RG, Peiris JS. Genesis of a highly pathogenic and potentially pandemic H5N1 influenza virus in eastern Asia. *Nature*. 2004; 430: 209-213.

[62]   Mukhtar MM, Rasool ST, Song D, Zhu C, Hao Q, Zhu Y, Wu J. Origin of highly pathogenic H5N1 avian influenza virus in China and genetic characterization of donor and recipient viruses. *J. Gen. Virol.* 2007; 88: 3094-3099.

[63]   Wu G, Yan SM. Randomness in the primary structure of protein: methods and implications. *Mol. Biol. Today*. 2002; 3: 55-69. http://www.horizonpress. com/mbt/v/v3/08.pdf.

[64]   Wu G, Yan S. Mutation trend of hemagglutinin of influenza A virus: a review from computational mutation viewpoint. *Acta. Pharmacol. Sin.* 2006; 27: 513-526.

[65]   Chutinimitkul S, Payungporn S, Chieochansin T, Suwannakarn K, Theamboonlers A, Poovorawan Y. The spread of avian influenza H5N1 virus; a pandemic threat to mankind. *J. Med. Assoc. Thai*. 2006; 89 Suppl 3: S218-S233.

[66] Chen JM, Chen JW, Dai JJ, Sun YX. A survey of human cases of H5N1 avian influenza reported by the WHO before June 2006 for infection control. *Am. J. Infect. Control.* 2007; 35: 467-469.

[67] Neumann G, Shinya K, Kawaoka Y. Molecular pathogenesis of H5N1 influenza virus infections. *Antivir. Ther.* 2007; 12: 617-626.

[68] Wu G, Yan SM. Analysis of distributions of amino acids, amino acid pairs and triplets in human insulin precursor and four variants from their occurrences according to the random mechanism. *J. Biochem. Mol. Biol. Biophys.* 2001; 5: 293-300.

[69] Wu G, Yan SM. Prediction of presence and absence of two- and three-amino-acid sequence of human monoamine oxidase B from its amino acid composition according to the random mechanism. *Biomol. Eng.* 2001; 18: 23-27.

[70] Wu G, Yan SM. Prediction of two- and three-amino-acid sequences of *Citrobacter Freundii* β-lactamase from its amino acid composition. *J. Mol. Microbiol. Biotechnol.* 2000; 2: 277-281. http://www.horizonpress.com/jmmb/v2/v2n3/05.pdf.

[71] Wu G, Yan SM. Random analysis of presence and absence of two- and three-amino-acid sequences and distributions of amino acids, two- and three-amino-acid sequences in bovine p53 protein. *Mol. Biol. Today* 2002; 3: 31-37. http://www.horizonpress.com/mbt/v/v3/05.pdf.

[72] Wu G, Yan S. Amino acid pairs susceptible to variants in human protein C precursor. *Protein Pept. Lett.* 2005; 10: 491-494.

[73] Wu G, Yan S. Determination of amino acid pairs in human p53 protein sensitive to mutations/variants by means of a random approach. *J. Mol. Model.* 2003; 9: 337-341.

[74] Bender C, Hall H, Huang J, Klimov A, Cox N, Hay A, Gregory V, Cameron K, Lim W, Subbarao K. Characterization of the surface proteins of influenza A (H5N1) viruses isolated from humans in 1997-1998. *Virology.* 1999; 254: 115-123.

[75] Gubareva LV, Novikov DV, Hayden FG. Assessment of hemagglutinin sequence heterogeneity during influenza virus transmission in families. *J. Infect. Dis.* 2002; 186: 1575-1581.

[76] Webster RG, Guan Y, Peiris M, Walker D, Krauss S, Zhou NN, Govorkova EA, Ellis TM, Dyrting KC, Sit T, Perez DR, Shortridge KF. Characterization of H5N1 influenza viruses that continue to circulate in geese in southeastern China. *J. Virol.* 2002; 76: 118-126.

[77] Spackman E, Senne DA, Davison S, Suarez DL. Sequence analysis of recent H7 avian influenza viruses associated with three different outbreaks in commercial poultry in the United States. *J. Virol.* 2003; 77: 13399-13402.

[78] Deem MW, Lee HY. Sequence space localization in the immune system response to vaccination and disease. *Phys. Rev. Lett.* 2003; 91: 068101-068104.

[79] Ferguson NM, Galvani AP, Bush RM. Ecological and immunological determinants of influenza evolution. *Nature.* 2003; 422: 428-433.

[80] Lin J, Andreasen V, Casagrandi R, Levin SA. Traveling waves in a model of influenza A drift. *J. Theor. Biol.* 2003; 222: 437-445.

[81] MathWorks Inc. *MatLab - The language of technical computing.* Version 6.1.0.450, release 12.1, 2001.

[82]   Wu G, Yan S. Timing of mutation in hemagglutinins from influenza A virus by means of unpredictable portion of amino-acid pair and fast Fourier transform. *Biochem. Biophys. Res. Commun.* 2005; 333: 70-78.

[83]   Wu G, Yan S. Timing of mutation in hemagglutinins from influenza A virus by means of amino-acid distribution rank and fast Fourier transform. *Protein. Pept. Lett.* 2006; 13: 143-148.

[84]   Tapping KF,. Mathias RG, Surkan DL. Pandemics and solar activity. *Can. J. Infect. Dis.* 2001; 12:1-12.

[85]   Wu G, Yan S. Searching of main cause leading to severe influenza A virus mutations and consequently to influenza pandemics/epidemics. *Am. J. Infect. Dis.* 2005; 1: 116-123. http://www.scipub.org/fulltext/ajid/ajid12116-123.pdf

[86]   Hagman M. Computer aided vaccine design. *Science.* 2000; 290: 80-82.

[87]   Bhasin M, Singh H, Raghava GPS. MHCBN: A comprehensive database of MHC binding and non-binding peptides. *Bioinformatics.* 2003; 19: 666-667.

[88]   Chou KC, Wei DQ, Zhong WZ. Binding mechanism of coronavirus main proteinase with ligands and its implication to drug design against SARS. *Biochem. Biophys. Res. Commun.* 2003; 308: 148-151.

[89]   Rost B. Review: protein secondary structure prediction continues to rise. *J. Struct. Biol.* 2001; 134: 204-218.

[90]   Liu J, Rost B. NORSp: Predictions of long regions without regular secondary structure. *Nucleic Acids Res.* 2003; 31: 3833-3835.

[91]   http://www.predictprotein.org/.

[92]   Bui HH, Sidney J, Peters B, Sathiamurthy M, Sinichi A, Purton KA, Mothé BR, Chisari FV, Watkins DI, Sette A. Automated generation and evaluation of specific MHC binding predictive tools: ARB matrix applications. *Immunogenetics.* 2005; 57: 304-314.

[93]   http://www.immuneepitope.org.

[94]   Larsen JE, Lund O, Nielsen M. Improved method for predicting linear B-cell epitopes. *Immunome. Res.* 2006; 2: 2.

[95]   http://www.cbs.dtu.dk/services/BepiPred.

[96]   Wu G, Yan SM. Theoretical analysis of drug treatment in haematological disease using Lanchester (Osipov) linear law. *Comp. Clin. Pathol.* 2002; 11: 113-118.

[97]   Wu G, Yan SM. Mathematical model of time needed for the immune system to detect and kill cancer cells in blood. *Comp. Clin. Pathol.* 2002; 11: 178-183.

[98]   Wu G, Yan S. Virus dynamics in vivo. *Am. J. Infect. Dis.* 2005; 1: 156-161. www.scipub.org/fulltext/ajid/ajid14156-161.pdf

[99]   Zambon MC Epidemiology and pathogenesis of influenza. *J. Antimicrob. Chemother.* 1999; 44(Suppl B): 3-9.

[100] Hochgürtel M, Kroth H, Piecha D, Hofmann MW, Nicolau C, Krause S, Schaaf O, Sonnenmoser G, Eliseev AV. Target-induced formation of neuraminidase inhibitors from in vitro virtual combinatorial libraries. *Proc. Natl. Acad. Sci. U. S. A.* 2002; 99: 3382-3387.

[101] Garman E, Laver G. Controlling influenza by inhibiting the virus's neuraminidase. *Curr. Drug. Targ.* 2004; 5: 119-136.

[102] von Itzstein M. The war against influenza: discovery and development of sialidase inhibitors. *Nat. Rev. Drug. Discov.* 2007; 6: 967-974.

[103] Nanni L, Lumini A. An ensemble of K-Local Hyperplane for predicting protein-protein interactions. *Bioinformatics.* 2006; 22: 1207-1210.

[104] Chou KC, Shen HB. Signal-CF: a subsite-coupled and window-fusing approach for predicting signal peptides. *Biochem. Biophys. Res. Commun.* 2007; 357: 633-640.

[105] Mdluli K, Ma Z. Mycobacterium tuberculosis DNA gyrase as target for drug discovery. *Infect. Disord. Drug. Targets.* 2007; 7: 159-168.

[106] Kontijevskis A., Wikberg JES, Komorowski J. Computational proteomics analysis of HIV-1 protease interactome. *Proteins: Struct. Funct. Bioinform.* 2007; 68: 305-312.

[107] Wu G. Frequency and Markov chain analysis of amino-acid sequence of human tumor necrosis factor. *Cancer. Lett.* 2000; 153: 145-150.

[108] Wu G. Frequency and Markov chain analysis of amino-acid sequence of human glutathione reductase. *Biochem. Biophys. Res. Commun.* 2000; 268: 823-826.

[109] Wu G. Frequency and Markov chain analysis of the amino acid sequence of human alcohol dehydrogenase α-chain. *Alcohol. Alcohol.* 2000; 35: 302-306. http://alcalc.oxfordjournals.org/cgi/reprint/35/3/302.

[110] Wu G. Frequency and Markov chain analysis of the amino-acid sequence of sheep p53 protein. *J. Biochem. Mol. Biol. Biophys.* 2000; 4: 179-185.

[111] Wu G. The first, second and third order Markov chain analysis on amino acids sequence of human tyrosine aminotransferase and its variant causing tyrosinemia type II. *Pediatr. Relat. Top.* 2000; 39: 37-47.

[112] Wu G, Yan SM. Frequency and Markov chain analysis of amino-acids sequence of human platelet-activating factor acetylhydrolase α-subunit and its variant causing the lissencephaly syndrome. *Pediatr. Relat. Top.* 2000; 39: 513-526.

[113] Wu G, Yan SM. Frequency and Markov chain analysis of amino-acid sequences of human connective tissue growth factor. *J. Mol. Model.* 2001; 5: 120-124.

[114] Feller W. *An introduction to probability theory and its applications.* 3rd edition. Vol, I. New York: Wiley; 1968; pp. 34-40.

[115] Tsuchiya Y, Kinoshita K, Nakamura H. Structure-based prediction of DNA-binding sites on proteins using the empirical preference of electrostatic potential and the shape of molecular surfaces. *Proteins.* 2004; 55: 885-894.

[116] Chou KC, Shen HB. Ensemble classifier for protein fold pattern recognition, *Bioinformatics.* 2006; 22: 1717-1722.

[117] Wu G, Yan SM. Analysis of distributions of amino acids in the primary structure of apoptosis regulator Bcl-2 family according to the random mechanism. *J. Biochem. Mol. Biol. Biophys.* 2002; 6: 407-414.

[118] Wu G, Yan SM. Analysis of distributions of amino acids in the primary structure of tumor suppressor p53 family according to the random mechanism. *J. Mol. Model.* 2002; 8: 191-198.

[119] Wu G, Yan S. Determination of sensitive positions to mutations in human p53 protein. *Biochem. Biophys. Res. Commun.* 2004; 321: 313-319.

[120] Gao N, Yan S, Wu G. Pattern of positions sensitive to mutations in human haemoglobin α-chain. *Protein Pept. Lett.* 2006; 13: 101-107.

[121] Skidgel RA, Erdös EG. Structure and function of human plasma carboxypeptidase N, the anaphylatoxin inactivator. *Int. Immunopharmacol.* 2007; 7: 1888-1899.

[122] Betakova T. M2 protein-a proton channel of influenza A virus. *Curr. Pharm. Des.* 2007; 13: 3231-3235.

[123] Wu G. Application of cross-impact analysis to the relationship between aldehyde dehydrogenase 2 and flushing. *Alcohol. Alcohol.* 2000; 35: 55-59. http://alcalc.oxfordjournals.org/cgi/reprint/35/1/55.

[124] Wu G, Yan S. Prediction of mutation trend in hemagglutinins and neuraminidases from influenza A viruses by means of cross-impact analysis. *Biochem. Biophys. Res. Commun.* 2005; 326: 475-482.

[125] Gordon TG. Hayward H. Initial experiments with the cross-impact matrix method of forecasting. *Futures.* 1968; 1: 100-116.

[126] Gordon TG. Cross-impact matrices – an illustration of their use for policy analysis. *Futures.* 1969; 2: 527-531.

[127] Enzer S. Delphi and cross-impact techniques: an effective combination for systematic futures analysis. *Futures.* 1970; 3: 48-61.

[128] Enzer S. Cross-impact techniques in technology assessment. *Futures.* 1972; 4: 30-51.

[129] Sage AP. *Methodology for large-scale systems.* New York: McGraw-Hill; 1977, pp. 165–203.

[130] Wikipedia, the free encyclopedia. Bayes' theorem. en.wikipedia.org /wiki /Bayes'_ theorem, 2007.

[131] Dayhoff MO, Schwartz RM, Orcutt BC. A model of evolutionary change in proteins, matrixes for detecting distant relationships. In: Dayhoff MO, editor. *Atlas of protein sequence and structure.* Vol 5, Washington, DC: National Biomedical Research Foundation; 1978; 345-358.

[132] Feng DF, Johnson MS, Doolittle RF. Aligning amino acid sequences: Comparison of commonly used methods. *J. Mol. Evol.* 1985; 21: 112-125.

[133] Karlin S, Ghandour G. Multiple-alphabet amino acid sequence comparisons of the immunoglobulin kappa-chain constant domain. *Proc. Natl. Acad. Sci. U. S. A.* 1985; 82: 8597-8601.

[134] Müller T, Spang R, Vingron M. Estimating amino acid substitution models: a comparison of Dayhoff's estimator, the resolvent approach and a maximum likelihood method. *Mol. Biol. Evol.* 2002; 19: 8-13.

[135] Wu G, Yan S. Determination of mutation trend in proteins by means of translation probability between RNA codes and mutated amino acids. *Biochem. Biophys. Res. Commun.* 2005; 337: 692-700.

[136] Wu G, Yan S. Determination of mutation trend in hemagglutinins by means of translation probability between RNA codons and mutated amino acids. *Protein. Pept. Lett.* 2006; 13: 601-609.

[137] Wu G, Yan S. Translation probability between RNA codons and translated amino acids, and its applications to protein mutations. In: Ostrovskiy MH, editor. *Leading-edge messenger RNA research communications.* New York: Nova Science Publishers; 2007; Chapter 3, pp. 47-65.

[138] Lee MS, Chen JS. Predicting antigenic variants of influenza A/H3N2 viruses. *Emerg. Infect. Dis*. 2004; 10: 1385-1390.

[139] Chi XS, Bolar TV, Zhao P, Tam JS, Rappaport R, Cheng SM. Molecular Evolution of Human Influenza A/H3N2 Virus in Asia and Europe from 2001 to 2003. *J. Clin. Microbiol*. 2005; 43: 6130-6132.

[140] Gramer MR, Lee JH, Choi YK, Goyal SM, Joo HS. Serologic and genetic characterization of North American H3N2 swine influenza A viruses. *Can. J. Vet. Res*. 2007; 71: 201-206.

[141] Nakajima K, Nobusawa E, Nagy A, Nakajima S. Accumulation of amino acid substitutions promotes irreversible structural changes in the hemagglutinin of human influenza AH3 virus during evolution. *J. Virol*. 2005; 79: 6472-6477.

[142] Carson ER, Cobelli C, Finkelstein L. *The mathematical modeling of metabolic and endocrine system: model formulation, identification, and validation*. New York: Wiley; 1983.

[143] Draper NR, Smith H. *Applied regression analysis*. 2nd edition. New York: Wiley; 1981.

[144] Hosmer DWJr, Lemeshow S. *Applied logistic regression*. 2nd edition. New York: Wiley; 2000.

[145] SPSS Inc. SigmaStat for windows version 3.00. 1992-2003.

[146] Healy MJR. Outliers in clinical chemistry quality-control schemes. *Clin. Chem*. 1979; 25: 675-677.

[147] Obenauer JC, Denson J, Mehta PK, Su X, Mukatira S, Finkelstein DB, Xu X, Wang J, Ma J, Fan Y, Rakestraw KM, Webster RG, Hoffmann E, Krauss S, Zheng J, Zhang Z, Naeve CW. Large-scale sequence analysis of avian influenza isolates. *Science*. 2006; 311: 1576-1580.

[148] Miller WG. OpenStat4. Version 13, Release 3, http://www.statpages.org /miller /openstat, 2006.

[149] Wu G, Yan S. Prediction of possible mutations in H5N1 hemagglutinins of influenza A virus by means of logistic regression. *Comp. Clin. Pathol*. 2006; 15: 255-261.

[150] Wu G, Yan S. Prediction of mutations in H5N1 hemagglutinins from influenza A virus *Protein. Pept. Lett*. 2006; 13: 971-976.

[151] Wu G, Yan S. Prediction of mutations engineered by randomness in H5N1 neuraminidases from influenza A virus. *Amino. Acids*. 2008; 34: 81-90.

[152] Wu G, Yan S. Prediction of mutations initiated by internal power in H3N2 hemagglutinins of influenza A virus from North America. *Int. J. Pept. Res. Ther*. 2008; 14: 41-51.

[153] Wu G, Yan S. Prediction of mutation in H3N2 hemagglutinins of influenza A virus from North America based on different datasets. *Protein. Pept. Lett*. 2008; 15: 144-152.

[154] Wu G, Yan S. Three sampling strategies to predict mutations in H5N1 hemagglutinins from influenza A virus. *Protein Pept Lett* (in press)

[155] Høybye JA. Model error propagation and data collection design. An application in water quality modeling. *Water, Air. Soil. Pollu*. 1998; 103: 101-109.

[156] Influenza virus resources. http://www.ncbi.nlm.nih.gov/genomes/FLU/Database /multiple.cgi, 2007.

[157] SPSS Inc. SigmaStat for Windows. Version 3.00, 1992-2003.

[158] SYSTAT Software Inc. Systat for Windows. Version 11.00.01, 2004.

[159] Jørfensen SF. *Fundamentals of ecological modeling*. 2nd edition. Amsterdam: Elsevier; 1994.

[160] Jacoby SLS, Kowalik JS. *Mathematical modeling with computer*. Englewood Cliffs, New Jersey: Prentice-Hall; 1980.

[161] Demuth H, Beale M. *Neural network toolbox for use with MATLAB*. Natick, MA: 2001.

[162] Hagan MT, Demuth HB, Beale MH. *Neural network design*. Boston, MA: PWS Publishing Company; 1996; Chapters 11 and 12.

[163] Wasserman PD. *Advanced methods in neural computing*. 1st edn, New York: Van Nostrand Reinhold; 1993; pp. 35-55.

[164] Wu G, Baraldo M, Pea F, Cossettini P, Furlanut M. Effects of different sampling strategies on predictions of blood cyclosporine concentrations in haematological patients with multidrug resistance by Bayesian and non-linear least squares methods. *Pharmacol. Res.* 1995; 32: 355-362.

[165] Wu G, Furlanut M. Prediction of blood cyclosporine concentrations in haematological patients with multidrug resistance by means of total, lean and different adipose dosing body weight using Bayesian and non-linear least squares methods. *Int. J. Clin. Pharmacol. Res.* 1996; 16: 89-97.

[166] Wu G, Cossettini P, Furlanut M. Prediction of blood cyclosporine concentrations in haematological patients with multidrug resistance by one-, two- and three-compartment models using Bayesian and non-linear least squares methods. *Pharmacol. Res.* 1996; 34: 47-57.

[167] Wu G, Pea F, Cossettini P, Furlanut M. Effect of the number of samples on Bayesian and non-linear least-squares individualization: a study of cyclosporin treatment of haematological patients with multidrug resistance. *J. Pharm. Pharmacol.* 1998; 50: 343-349.

[168] Wu G, Furlanut M. Prediction of serum vancomycin concentrations using one-, two-, and three-compartment models with the implemented population pharmacokinetic parameters and with the Bayesian method. *J. Pharm. Pharmacol.* 1998; 50: 851-856.

[169] Kohonen T. *Self-Organization and associative memory*, 2nd Edition, Berlin: Springer-Verlag, 1987.

[170] Wu G, Yan S. Prediction of mutations in H1 neuraminidases from North America influenza A virus engineered by internal randomness. *Mol. Divers.* 2007; 11: 131-140.

[171] Wu G, Yan S. Prediction of mutations engineered by randomness in H5N1 hemagglutinins of influenza A virus. *Amino. Acids.* 2008; (in press).

[172] Scientific Consulting. *WinNonlin user's guide*. North Carolina; 1995.

[173] en.wikipedia.org/wiki/Mean_squared_error and en.wikipedia.org /wiki /Mean_ absolute_error.

[174] Wu G, Yan S. Improvement of prediction of mutation positions in H5N1 hemagglutinins of influenza A virus using neural network with distinguishing of arginine, leucine and serine. *Protein. Pept. Lett.* 2007; 14: 465-470.

[175] Garbrielsson J, Weiner D. *Pharmacokinetic and pharmacodynamic data analysis: concepts and applications*. Stockholm: Swedish Pharmaceutical Press; 1994.

[176] Notari RE. Biopharmaceutics and clinical pharmacokinetics. 4th edition, New York: Dekker; 1987.

[177] Walpole RJ, Myers RH. *Probability and statistics for engineers and scientists*. 4th edition. New York: Macmillan Publishing Company. 1989.

[178] Wu G. Squared correlation coefficient of measured values versus predicted values in linear and monoexponential regressions. *Eur. J. Drug. Metab. Pharmacokinet.* 2002; 27: 113-117.

[179] Wu G. An extremely strange observation on the equations for calculation of correlation coefficient. *Eur. J. Drug. Metab. Pharmacokinet.* 2003; 28: 85-92.

[180] Holman JP. *Heat transfer*. 7th edition. New York: McGraw-Hill Company, 1990.

[181] Langhaar HL. Dimensional analysis and theory of models. New York: Wiley, 1967.

[182] Wu G, Yan S. Improvement of model for prediction of hemagglutinin mutations in H5N1 influenza viruses with distinguishing of arginine, leucine and serine. *Protein. Pept. Lett.* 2007; 14: 191-196.

[183] Ahn I, Son HS. Comparative study of the hemagglutinin and neuraminidase genes of influenza A virus H3N2, H9N2, and H5N1 subtypes using bioinformatics techniques. *Can. J. Microbiol.* 2007; 53: 830-839.

Index